Contributions of Physiology to the Understanding of Diabetes

Springer-Verlag Berlin Heidelberg GmbH

Albert E. Renold 1923–1988

G.R. Zahnd   C.B. Wollheim   (Eds.)

Contributions of Physiology to the Understanding of Diabetes

Ten Essays in Memory of Albert E. Renold

With 38 Figures and 7 Tables

 Springer

Dr. GASTON R. ZAHND
Fondation pour Recherches Médicales
Faculté de Médecine
Université de Genève
64, Avenue de la Roseraie
1211 Genève 4
Switzerland

Professor CLAES B. WOLLHEIM
Division de Biochimie clinique
et de Diabétologie expérimentale
Centre Médical Universitaire
1211 Genève 4
Switzerland

ISBN 978-3-642-64422-1   ISBN 978-3-642-60475-1 (eBook)
DOI 10.1007/ 978-3-642-60475-1

Library of Congress Cataloging-in-Publication Data.   Contributions of physiology to the understanding of diabetes: ten essays in memory of Albert E. Renold/[edited by] R.G. Zahnd, C.B. Wollheim.   p.   cm.   Includes bibliographi- cal references and index. ISBN 978-3-642-64422-1(hardcover: alk. paper) 1. Diabetes – Pathophysiology. 2. Renold, A.E. (Albert Ernst), 1923– . I. Zahnd, G.R. (Gaston R.), 1925– . II. Wollheim, C.B. (Claes B.), 1943– . RC660.C67 1997 616.4'6207 – dc21 96-29827

© Springer-Verlag Berlin Heidelberg 1997
Softcover reprint of the hardcover 1st edition 1997

Cover design: Design & Production GmbH, Heidelberg

Typesetting: Best-set Typesetter Ltd., Hong Kong

SPIN: 10500858   23/3134/SPS – 5 4 3 2 1 0 – Printed on acid-free paper

Foreword

This book is dedicated to Albert E. Renold in honour of his profound commitment and outstanding contributions to the development of diabetes research. The need for such a monograph was brought home to us many times as we discussed experimental approaches – some new, some old – to the study of diabetes and the still missing link between physiological variations and overt impairment of insulin secretion and sensitivity. Many of us had the good fortune to know Albert Renold and were privileged to share the excitement of his research. For these reasons, scientific colleagues and friends decided to bring together their thoughts on major topics which, in part, have been seeded, grown or matured in the remarkable "culture medium" Albert knew how to prepare so well.

In the summer of 1948, a freshly graduated Swiss M.D. disembarked in Boston to begin a research fellowship with Elliott P. Joslin and Alexander Marble at the New England Deaconess Hospital. What might have been a single migratory event turned out to become the cornerstone of Albert Renold's postdoctoral training years, learning to cross boundaries of medicine, endocrinology and biological chemistry under the further guidance of George W. Thorn, A. Baird Hastings and associates. By this time the gifted investigator had become fully acquainted with Rachmiel Levine's quest into the mode of action of insulin. Albert always liked to emphasize the crucial influence his distinguished mentors exerted on him during the 1950s. After a period of fundamental work at Harvard's carbohydrate research laboratory he moved to the Baker Clinic research laboratory, later renamed for Dr. Joslin. This innovative laboratory soon became a diabetes centre attracting young physicians and scientists to explore experimental and clinical intermediary metabolism, adipose tissue physiology, experimental insulitis, and the pathogenesis of diabetes in man and animals.

Facing the challenge to extend his activities under as yet less advanced conditions in the "Old World," Renold accepted the call from the Medical Faculty of the University of Geneva, where he started a new life in keeping with the spirit he had acquired in Boston. From 1963 the Institut de Biochimie Clinique (IBC) was a communicative continuum of scientific interaction touching on many of his previous interests: insulin and glucagon biosynthesis and secretion, metabolic, vascular and neurological diseases. He was convinced that a better understanding of the pathogenesis and prevention of diabetes would surely result from observations in animals with either genetically or experimentally induced diabetic syndromes. Inspired by Albert's drive and enthusiasm, his co-workers at the IBC rapidly established contacts with other teams, both locally and extra muros. The studies on the endocrine pancreas led to a particularly fruitful partnership with Charles Rouiller, Lelio Orci and associates from the department of morphology.

Albert Renold's capacity to dialogue and to listen was instrumental in the founding of the European Association for the Study of Diabetes. Clearly, neither the Renaissance surroundings in Tuscany at Monte Catini, where the first annual meeting of the EASD took place, nor the setting at the seaside of Scheveningen in Holland, where Renold assisted Alex Muller in constituting the European Society for Clinical Investigation, was a matter of hazard. Both initiatives were essential forerunners of a gradually developing international scientific community in Europe.

For practical purposes the number of contributors to the present volume has had to be limited. In order to compensate for inevitable omissions, the names of close collaborators appear in the Selected Bibliography of Albert Renold added as an appendix. The book is so arranged that some essays set the background for what follows.

The first section deals with general aspects of cellular biochemistry and metabolism. The reader will enjoy Martin Rodbell's reminiscences about a sabbatical leave in Geneva at the time of his prime work on a new theory of hormone action through the regulation of membrane signal transduction by GTP-binding proteins. In the next chapter the particular area of osmolite metabolism is addressed by Louis Hue, proposing physiological mediators in the activation of glycogen synthesis and lipogenesis by swelling liver cells.

In the following section four related chapters focus on the pancreatic β-cell and islets of Langerhans. Paolo Meda reveals how an intriguing family of transmembrane proteins, called connexins, is involved in cell-to-cell communication and why defects of cell coupling may be pathogenetic in β-cell dysfunction. Recent data about the β-cell specific expression of the insulin gene and its regulation by glucose and cAMP second messengers are discussed by Jacques Philippe, while Franz Matschinsky provides the biochemical foundation of fuel sensing by the pancreatic islets, bringing further evidence for the essential glucose sensor role of glucokinase in the β-cell. Current knowledge of the modulation of the exocytotic process in insulin secretion by Ca^{2+} influx with the regulatory function of submembrane and cytosolic proteins is reviewed by Claes Wollheim. These studies illustrate the elegant use of advanced techniques of cellular and molecular biology in experiments on insulin-secreting cell lines.

A third section pertains to the action of the insulin and insulin-like growth factors at the level of their target cells, as well as to the tissue-specific regulation of glucose transport systems. An additional chapter deals with conceptual aspects of pathogenetic mechanisms leading to insulin-dependent diabetes. Pierre De Meyts has proposed a binding model of insulin and IGF-1 to the subunit structure of their allosteric receptors. The author's present contribution provides an updated compilation of emerging signalling ways and byways along which receptor kinase activities may promote metabolic and mitogenic effects of insulin and other growth factors. Not surprisingly, attempts are currently being made by engineering polygenic mice to find the way through the labyrinth of potential postreceptor defects in NIDDM. One of the most significant final effects of insulin action consists in controlling the transport of glucose into the cells. Sam Cushman tells the revisited story of mechanisms underlying the glucose transporter process. The interplay of insulin, growth hormone, insulin-like growth factors and nutrition which governs overall energy metabolism is described by Ruedi Froesch. This overview covers evolutionary aspects of the present somatomedin concept as well as clinical trials in which recombinant human IGF-1 is used as an adjuvant to insulin therapy in selected groups of diabetic patients.

Lastly, Jorn Nerup and coworkers describe two distinct phases of immune effector events and cytokine-mediated phenomena which are implicated in the selective destruction of pancreatic β-cells in insulin-dependent diabetes mellitus (IDDM). Recent evidence, derived from observations in animal models, provides insight into the polygenic control of the protective or deleterious mechanisms in human diabetes that can be deduced from epidemiological studies.

Each of the essays illuminates a trajectory of the disease from the point of view of the author's special observatory. Not only is the diabetic disorder an important source of knowledge about normal physiology, but what we learn about pathophysiology in turn furthers the quality of medical practice of diabetes. With this in mind, Albert Renold attempted to integrate biological science with the permanent evolution of knowledge. As a prominent leader, he went beyond the high expectations of Dr. Joslin, who, at the age of 90 years, conveyed the following leitmotif by way of a farewell note to some other exchange visitor:

Geneva, Switzerland   Gaston R. Zahnd

Preface

I am honored to write the foreword for a book dedicated to a special individual, but it is nearly impossible to put into words the many and varied aspects of Albert Renold. The authors of this volume are all outstanding contributors to diabetes research and each has been recognized by his peers in exceptional ways, including a Nobel Prize. That this distinguished group, representing such diverse interests, would set aside time to honor another scientist captures the truly remarkable nature of Albert. Each one of the authors has a personal connection to him, and if asked, I am sure each would feel that he had a unique relationship with him and owes a debt of gratitude that could never be adequately repaid. I count myself among them, because, not only did Albert understand and dedicate himself to new knowledge, he understood and dedicated himself to scientists. He was sensitive to their personal needs as well as to their professional goals.

Those of us who were fortunate enough to spend time at the Institut de Biochimie Clinique share a bond which truly represents a shared vision of what science and scientific inquiry should be and how research and living ought to be integrated. The impact that Albert Renold had was so great that the European Association for the Study of Diabetes has named a fellowship after him, and the American Diabetes Association now awards the Albert Renold medal each year to an outstanding mentor, a unique recognition of the international impact that he had and the particular contribution he made to training.

Despite all his success, he was not immune to struggle, disappointment, and tragedy. Nevertheless, his interests in others always came first, and support of his coworkers, collaborators, students, fellows and faculty (both permanent and visiting) was never wanting. His command of language was legendary. His family relationships were strong and vital parts of the Institut's success. Everyone contributed because they wanted to participate in an exciting, stimulating, and challenging experience. We still miss the Saturday laboratory sessions, the group outings, the philosophical discussions, the administrative musings, and the everyday give-and-take with a gifted scientific colleague. Fortunately, under Claes Wollheim's expert leadership, the Institut continues to thrive.

The papers presented here represent reviews of important areas of diabetes research today. Albert Renold contributed to all of them, including insulin secretion and beta-cell function, fat cell metabolism and insulin action, and immunity in diabetes. He would no doubt be interested and pleased to read this volume, although he would probably be embarrassed by its dedication. Why did we wait so long to publish this work? I'm not sure anyone can answer that question; perhaps it's because it takes time to adjust to a loss of such magnitude. This volume, then, is a sign of our acceptance of a passing, our continuing pleasure with the progress of science, and Albert Renold's many achievements. I am sure that Albert would want to congratulate and encourage all of the contributors and wish the reader a stimulating and challenging reading experience.

Seattle, WA, USA   DANIEL PORTE JR.

Acknowledgements

The editors would like to express their great respect and appreciation to those who deserve particular mention for their help in the preparation of this book, thus carrying on the flame that A.E.R. transmitted to his loyal staff members: Mrs. Claire-Lise Moriaud Noverraz, administrator, provided original documents and secretarial assistance; Mrs. Jean Gunn, librarian, co-edited the manuscripts.

The photograph of Professor Albert Renold is reproduced with kind permission of H.W. Stricker. The cover illustration is taken from a pen and walnut stain drawing of the Institut de Biochimie Clinique, Sentier de la Roseraie (1982) reproduced with kind permission of the artist, Henri Noverraz.

The publication of the monograph was made possible by a generous grant-in-aid from the Institute for Human Genetics and Biochemistry, Geneva, Switzerland.

Geneva, Switzerland   GASTON R. ZAHND
                                                             CLAES B. WOLLHEIM

Contents

Foreword
G.R. Zahnd V

Preface
D. Porte Jr. IX

Acknowledgements XI

Thoughts on the Regulation of Adenylyl Cyclase Systems
by Hormone-Sensitive G-Proteins: A Tribute to Albert E. Renold
M. Rodbell 1

Anabolic Response to Cell Swelling in the Liver
L. Hue, V. Gaussin, and U. Krause 10

Intercellular Communication and Insulin Secretion
P. Meda 24

Expression of the Insulin Gene and Its Regulation
J. Philippe 43

Implications of the Glucokinase Glucose Sensor Paradigm
for Pancreatic β-Cell Function
F.M. Matschinsky 54

How Ca^{2+} and Other Signalling Pathways Control the Exocytosis
of Insulin in the β-Cell
C.B. Wollheim 68

The Mechanism of Insulin Receptor Binding, Activation
and Signal Transduction
P. De Meyts and K. Seedorf 89

Cell Biology of Insulin Action on Glucose Transport: Looking Back
S.W. Cushman 108

Insulin-like Growth Factor: Endocrine and Autocrine/Paracrine
Implications and Relations to Diabetes Mellitus
E.R. Froesch 127

On the Pathogenesis of Insulin-Dependent Diabetes Mellitus in Man:
A Paradigm in Transition
J. NERUP, T. MANDRUP-POULSEN, F. POCIOT, A.E. KARLSEN,
H.U. ANDERSEN, U.B. CHRISTENSEN, T. SPARRE, J. JOHANNESEN,
and O.P. KRISTENSEN 148

Selected References from the Bibliography of Albert E. Renold, M.D. 160

Subject Index 169

List of Contributors

SAMUEL W. CUSHMAN
Experimental Diabetes, Metabolism, and Nutrition Section, Diabetes Branch,
National Institute of Diabetes and Digestive and Kidney Diseases,
National Institutes of Health, Building 10, Room 5N102, 10 Center Dr MSC 1420,
Bethesda, MD 20892–1420, USA

PIERRE DE MEYTS and COLLABORATOR
Hagedorn Research Institute, Niels Steensens Vej 6, 2820 Gentofte, Denmark

ERNST RUDOLF FROESCH
Division of Endocrinology and Metabolism, Department of Internal Medicine,
University Hospital, 8091 Zürich, Switzerland

LOUIS HUE and COLLABORATORS
Hormone and Metabolic Research Unit, ICP-UCL 75.29, Avenue Hippocrate 75,
1200 Brussels, Belgium

FRANZ M. MATSCHINSKY
Diabetes Research Center, University of Pennsylvania, School of Medicine,
36th and Hamilton Walk, Philadelphia, PA 19104-6015, USA

PAOLO MEDA
Department of Morphology, University of Geneva, Centre Médical Universitaire,
1211 Geneva 4, Switzerland

JØRN NERUP and COLLABORATORS
Steno Diabetes Center, Niels Steensens Vej 2, 2820 Gentofte, Denmark

JACQUES PHILIPPE
Unité de Diabétologie Clinique, Hôpital Cantonal Universitaire, 1211 Geneva 4,
Switzerland

MARTIN RODBELL
National Institute of Environmental Health Sciences, Research Triangle Park,
NC 27709, USA

CLAES B. WOLLHEIM
Division de Biochimie clinique et de Diabétologie expérimentale,
Centre Médical Universitaire, 1211 Geneva 4, Switzerland

Thoughts on the Regulation of Adenylyl Cyclase Systems by Hormone-Sensitive G-Proteins: A Tribute to Albert E. Renold

M. RODBELL

Several years ago I witnessed in New York City a play called "Six Degrees of Separa-tion." Its underlying theme dealt with the associations between people that are often not recognized until some circumstance, as if by magic, reveals a web-like interaction between unsuspecting subjects. I recall being stimulated by this phenomenon to the point that I began to think about my own "six degrees of separation" with others. One individual stood out in this paradigm – Albert E. Renold. In many ways I and many other scientists who knew him were influenced by the extraordinary character of this exuberant, cultured, and humane man.

In the early 1960s I had known of Albert Renold through his role as co-editor with George Cahill of an edition of the *Handbook of Physiology* dealing with adipose tissue. I was invited to submit a chapter on the metabolic behavior of isolated adipocytes because of my discovery that treatment of adipose tissue with a crude collagenase preparation resulted in disaggregation of the tissue matrix and release of the adipose cells. These cells were easily separated from stroma and blood vessels by simply spinning the fat-laden cells to the surface, leaving the other more dense cells to sediment. With this very simple technique I determined that the adipocyte contains lipoprotein lipase, a finding that was the original reason for my attempts to isolate adipocytes. Others later showed that the adipocyte is a major source of the enzyme in its passage to the endothelium of blood vessels where it performs its function of catalyzing the hydrolysis of chylomicron triglycerides. But that is another story. At the point of isolating the cells I was faced with the question, posed by Professor Bernardo Houssay while visiting my laboratory, of whether the isolated cells are functional, particularly with respect to the known effects of insulin on glucose uptake and me-tabolism. That question was my first point of contact with the subject of how hor-mones such as insulin act. My linkage to the endocrine world had begun when the adipocyte displayed extraordinary sensitivity to insulin. The web with Albert was now in its formative stage.

Shortly after my findings with insulin, Robert Williams, a noted endocrinologist at the University of Washington in Seattle (where I had received my Ph.D.), entered the unsuspected web by telling me of his experiences and knowledge gained from inves-tigating insulin action: "Be gentle with the cells in your investigations with insulin; crush them in the usual biochemist's fashion and you will surely find that the hor-mone no longer works." In hindsight, this admonition was not fully justified. Years later it was discovered that insulin and many growth-promoting hormones act on isolated purified receptors by stimulating the phosphorylation of tyrosine groups in protein substrates. However, beyond that initial stage, it is still not clear how insulin promotes its metabolic and glucose-transport actions. In 1964, I took Bob Williams' advice seriously. In the initial stage I treated the cells with purified phospholipase A and C from snake venom and clostridia, the idea being that their hydrolytic activity might selectively and "gently" open up selective regions of the cell membrane, releasing fat and soluble enzymes while retaining the insulin-sensitive machinery necessary for insulin's effects on glucose uptake and metabolism. To my great surprise, the treated cells remained intact and behaved as if they were exposed to insulin. The production of $^{14}CO_2$ from glucose-1-^{14}C rose to levels observed with insulin-treated cells. The insulin-mimetic actions of the phospholipases were also observed in protein synthesis. None of the treated cells showed any obvious changes in structural integrity. I hypothesized from these findings that the surface membrane phospholipids may have specialized structures which insulin, through its receptor, and phospholipases can alter (the latter irreversibly) to affect glucose transport and utilization, amino acid transport and utilization (protein synthesis), and inhibition of lipolysis induced by lipolytic hormones. Certainly such far-fetched speculations were not taken seriously at that time. In fact, phospholipids were deemed essential to maintain the necessary "fluidity" that allows embedded proteins to move freely so that interactions between receptors and effectors can take place. Thus, the "mobile receptor theory" gave much prominence to the actions of the insulin receptor. This theory, still tenable to many scientists, no longer has the prominence it had during the 1970s and 1980s. Membrane phospholipids are now viewed as complex structures that can form very compact domains depending on the charge and structure of the head groups, the degree of unsaturation of the fatty acids at the sn2 position, and the type of saturated fatty acids at the sn1 position [1]. The degree of compactness determines the packing domains and the ability of intrinsic and associated membrane proteins to interact with the membrane lipid matrix. One can reasonably speculate, therefore, that the ability of proteins such as receptors and regulatory elements to associate is the result of the ordering of membrane lipids in the membrane bilayer and not solely or even at all on the fluidity of the membrane lipids. Hence, perhaps the basis of the insulin-mimetic actions of phospholipases is change in the packing structure. Certainly, this idea merits investigation, particularly in the light of the newly revealed structures of membrane lipids.

Still the reductionist biochemist, I was determined in the 1960s to reduce the complexity of the adipocyte insulin system. Knowing that the lipolytic process can lead to altered hormonal responses in this cell, I began a different approach utilizing treatment with hypotonic media. This osmolytic procedure ruptured the cell membrane allowing release of the main large fat droplet, the nucleus in the majority of cells, and "soluble" cytosolic enzymes. Mitochondria could be seen in some of the adipocyte "ghosts." Most importantly, these "ghost" preparations contained several of the receptors for hormones known to stimulate lipolysis via the production of cyclic 3'5'-AMP (cyclic AMP). Insulin displayed a modest stimulation of glucose oxidation in "ghosts" whereas the lipolytic hormones elicited marked stimulation of cyclic AMP. Soon after publication of these findings, I received an invitation from Albert Renold to present a seminar at the Institut de Biochimie Clinique in Geneva. The web began to tighten. Insulin, adipocytes, and lipolytic hormones had converged (perhaps conspired) to bring us together. I felt instantly at home with the atmosphere in the Institut, the surrounding residential nature of its setting, and the casual demeanor of its occupants. Overseeing this almost bucolic (without the wine) atmosphere hovered the warm personality of Albert with his unbelievable ability to display the open-minded behavior of an American combined with the charm and erudition of a cultivated European.

One year later arrived my second post-doctoral fellow from Argentina. Lutz Birnbaumer began his now historical efforts to determine the kinetic parameters responsible for the stimulatory effects of lipolytic hormones on adenylate (now adenylyl) cyclase activity. During this time I received a telephone call from Albert: "Would you be willing to take charge of the Institut during my sabbatical year in Seattle with Robert Williams? My colleagues took a vote and unanimously suggested you as my temporary replacement. You can have our apartment in the Cours des Bastions." The web now included Robert Williams and Albert Renold. My wife and four children easily settled into a year they will never forget. Already accustomed to European life from their year in The Hague behind the dunes lining the North Sea, 7 years later they experienced the many facets of living in the multi-cultured atmosphere of Geneva with its combination of status-seeking Americans, United Nations bureaucrats, the authentic Genevois with their pseudo-aristocratic bearing, the literally underground culture of the Cern physicists, and the academic scientists at the University of Geneva. The Institut proved similarly constituted with different people sorting into distinct but related categories of experimentation. Torben Clausen arrived from Aarhus and teamed up with me to investigate the effects of hormones on ion and amino acid transport in adipocyte ghosts. Another group chose to carry out similar experimentation with intact adipocytes. Another group investigated the effects of glucose on insulin release from perfused rat pancreas. The Institut became a hot-house of enthusiasts in their specialties, each vying for prominence, a place in the sun. A National Institutes of Health-style of competitiveness arose occasionally with heated interactions. By concentrating on research I had lost control. Albert arrived, almost too late, to the rescue. Immediately the atmosphere of the Institut became more relaxed, the scientists less frenetic in their experimental pursuits, and Torben and I managed to publish two interesting papers on the transport properties of adipocyte ghosts. Alas, no hormonal effects on the transport processes were consistently observed.

Another important character in the web was Lelio Orci. Lelio had only recently arrived in Geneva as a very young novice in electron microscopy. Scarcely able to speak either French or English, Lelio easily communicated with me through his excellent photos and his Italian hand-waving and gestures. Both Albert and I recognized his extraordinary talents even at that early stage of his career. Today he is considered a giant in his field, having contributed enormously to our knowledge of membrane-trafficking and secretory processing. At that time he was the source of uninhibitable energy and enthusiasm; I relished his visits to the Institut.

On the other side of the Atlantic, Lutz was churning out great quantities of kinetic data. Without telefax and e-mail we managed to share our thoughts in a reciprocal fashion, the distance between us providing for needed time for reflection. With Albert home again with his wonderful wife, Jacqueline, and their sons Frederic and Marc-Andre filled with stories of great adventures in the Wild West, our families finally were able to establish a relationship that lasted until the tragic end to the lives of Albert and Jacqueline. In quiet moments I discussed with Albert the possible significance of Lutz's work. Lutz and I were beginning to see that the concept developed by

Sutherland that the adenylate cyclase system is composed of two elements, receptor and enzyme, might not be correct. With the seed of doubt planted and the Geneva experience at its end, I returned to the National Institutes of Health with the prospect of unearthing a new concept.

The multi-receptor adenylate cyclase system in the rat adipocyte proved to be a "mother-lode" of information. Within 3 short years we discovered that this complex system contains a regulatory site for Mg^{2+}, that receptors for five lipolytic hormones are distinct entitites, each appearing to act on a common enzyme, and that fluoride ion, which stimulated the enzyme, could also inhibit the enzyme. Given also the complex kinetic effects of ATP on cyclic AMP production, it also appeared that the system contains a separate regulatory site for ATP. Once again phospholipids appeared to have a decisive role in the ability of hormones to activate the enzyme. Enriched with new members (Steve Pohl, Michiel Krans) we fashioned the idea that there must be some independent component intervening between receptor and enzyme. I coined this element "transducer" under the influence of Norbert Wiener and the guidance of Oscar Hechter. A new web of interacting individuals formed during the ensuing decade which established that the transducer is a site both for Mg and for guanosine triphosphate (GTP) in both the activation and the inhibition of adenylate cyclase activity. With two distinct GTP-processes regulating the enzyme independently and with distinctly different properties we had no difficulty in ascertaining that both processes are carried out by GTP-binding proteins that mediate the transfer of "information" between receptors and enzyme. Within a decade such GTP-binding proteins were found in all living cells except for bacteria and are the principal mediators for the actions of hundreds of external signals ranging from photons to protein hormones. The stage was set for understanding the fundamental nature of the transduction process, i.e., the puzzle of how three components (in the case of adenylate cyclase systems) are capable of exchanging information when two (receptor and enzyme) are embedded in the matrix or lipid envelope of the surface membrane and the transducer is somehow joined to the interior or cytoplasmic side of the membrane.

I have reviewed [2] most of the thought processes and the experiments that led to the idea that receptors are linked to oligomeric forms of GTP-binding proteins (termed G-proteins). I had concluded from our studies during the 1970s that the structure/regulatory functions of the G-proteins when attached to receptors are decidedly different from these functions when G-proteins activate adenylyl cyclase. Based on target or irradiation analysis of the functional masses of adenylyl cyclase systems in both rat hepatocytes and adipocytes (both stimulatory and inhibitory forms), it seemed likely that the structures of G-proteins combined with receptors were oligomeric, whereas when the G-proteins are activated by their respective receptor types and combine with effectors such as adenylyl cyclase they are likely to be monomeric [3]. Cross-linking of G-proteins in their native membrane environment supported the oligomeric structure [4]. Detergent studies, combined with hydrodynamic studies, supported the contention that activation of G-proteins by guanine nucleotides (induced by hormone binding to receptor) leads to release of monomers from oligomers [5].

During the 1980s great strides had been made in deciphering the structures of G-proteins, culminating in the realization that they are heterotrimeric structures composed of α subunits that bind and degrade GTP to guanosine diphosphate (GDP) and a tightly linked $\beta\gamma$ complex that is both necessary for hormonal activation of the α subunit (by inducing exchange of bound GDP with GTP) and for attachment of this unit to the receptor and membrane at its cytoplasmic surface. When several investigators found that non-hydrolyzable analogs of GTP (GTPγS, Gpp(NH)p) and fluoride (complexed with aluminum or magnesium ions) were capable of inducing dissociation of the heterotrimer into free α and the $\beta\gamma$ complex, the predominant theory postulated that dissociation was the fundamental mechanism for the pleiotropic mechanisms by which hormones stimulate two or more independent processes. The relative simplicity of this theory and the revelation that the α subunits act independently of the $\beta\gamma$ complexes conspired to convince almost everyone of its validity. I was not convinced for two fundamental reasons: (1) GTP, the natural activator, was virtually ineffective as a dissociative agent, and (2) dissociation with the analogs occurred in detergents that extract heterotrimeric but not oligomeric forms of G-proteins [5]. These findings suggested the necessity of a better, albeit a more complex hypothesis, which is illustrated schematically in Fig. 1.

Based in large measure on the striking similarities in the regulation of oligomeric structures of actin and tubulin by adenine and guanine nucleotides, respectively, with that of G-proteins in their oligomeric state, this model incorporates the idea of "dynamic instability". Actin oligomers, for example, are asymmetric structures having at one end monomers containing ADP that are released, whereas the other barbed end has an affinity for taking up monomers containing ATP. Hence, in a "teeter-totter" fashion an oligomer can display both release and uptake of monomers depending on the distribution and type of adenine nucleotide situated along the oligomeric chain. Myosin, with its attachments (strong and weak forces) to the oligomeric actin, induces exchange of ADP and ATP during its movement along the chain, allowing for a change in the distribution of these nucleotides (and inorganic phos-

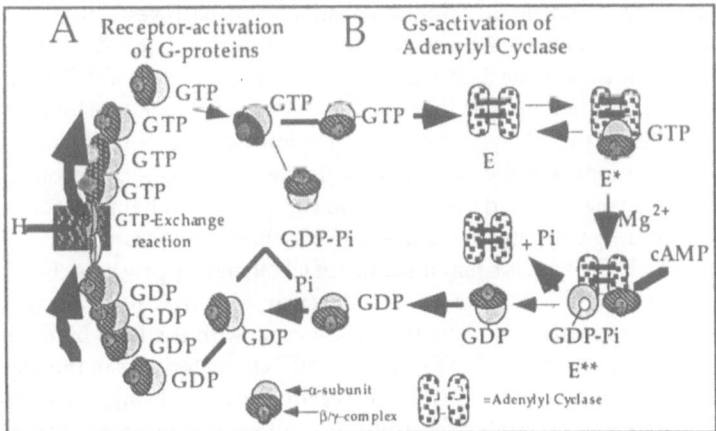

Fig. 1A,B. Regulation of adenylyl cyclase by hormones and guanine nucleotides. A two-cycle model. For details, see text. *Gs*, G-proteins; *cAMP*, cyclic adenosine monophosphate; *Pi*, inorganic phosphate; *E*, enzyme; *asterisks*, different activity states

phate bound to specific groups) resulting in the dynamic changes in actin that, in muscle, bring about the contractile forces necessary for muscle function. When ATP and its products are not available, actin and myosin form strong interactions resulting in "rigor".

It took very little imagination to visualize similar interactions of receptors and oligomeric G-proteins, the major differences being that the receptors are intertwined with the membrane (in a serpentine seven-domain spanning fashion) such that their cytoplasmic surfaces interact with G-protein monomers as they migrate, in pulsatile fashion, along the chain. This is illustrated in Fig. 1A. Each contact of the receptor with its selective G-protein monomer brings about an exchange of bound GDP with exogenous GTP; the hormone-induced structural changes between receptor and G-protein (accompanied by the nucleotide exchange reaction) maintain the reversible forces between receptor and G-proteins, thus allowing for movement of the receptor along the chain. If there are no guanine nucleotides available (bound or free), the attractive forces between bound hormone and receptor coupling with G-protein become pseudo-irreversible, as was first evidenced in the earliest study of glucagon binding to its receptor in rat liver membranes [6]. The binding of either GTP or GDP to the G-protein releases the strong-coupling forces resulting in dissociation of hormone from receptor and continued movement of the receptor along the chain. Thus, this model accounts for the apparent paradox that GTP and GDP equally affect the dissociation rate constant for hormone binding whereas, in the case of the effectors such as adenylyl cyclase, GTP activates whereas GDP inhibits, a phenomenon explained below. The tight-binding state of the receptor G-protein complex, reflected by the irreversible binding of hormones in the absence of guanine nucleotides, is possibly the most important state of the receptor for hormone action. Formed by hormone binding only when the coupled G-protein is unoccupied by GTP or GDP, its pseudo-irreversible state (likened to the "rigor" state of myson-actin in the absence of adenine nucleotides) suggests it is the high-affinity state of the receptor required for physiological action. This state induces the GDP/GTP exchange reaction and must occur transiently at the moment when the nucleotide binding site(s) are vacant. In isolated membranes, hormone binding constants can be investigated by manipulating the presence of GTP or GDP and should be included in determining the hormone binding constants during the activation of G-protein (and effectors). To my knowledge, none of the published models that attempt to correlate hormone binding and action takes this state of the adenylyl cyclase system into account. The obvious difficulties in such computations are the transient nature of the receptor/G-protein state during the exchange reaction and the problem of completely removing all traces of bound GTP or GDP during experimental studies with isolated membranes.

GTP hydrolysis is a major feature of all regulatory processes involving GTP binding proteins. In the case of small molecular weight G-proteins (RAS, for example), GTPase-activating proteins (GAP) associate with the GTP binding proteins and are thought to be responsible for the "on-off" characteristics of the regulatory processes; i.e., GTP binding induces the "on" reaction, GTPase brings about the "off" reactions. A simple and certainly powerful means of explaining regulation, GTPase regulation has become a prime theory in the regulatory behavior of all G-protein mediated processes, including those involving hormonally regulated heterotrimeric proteins. It should be emphasized, however, that high levels of GTPase activity generally result from interactions of G-proteins with GAP proteins or, in the case of heterotrimeric G-proteins, with their effectors as indicated by recent studies with hormonally activated phospholipases [7]. As mentioned previously, GTPase activity is minimal when studied with purified heterotrimeric G-proteins.

Another important aspect of G-protein regulation of effector systems such as adenylyl cyclase is that there are now at least ten different species of adenylyl cyclase, each having distinct regulatory properties with respect to activation or inhibition by calcium ions, $\beta\gamma$ complexes, and stimulation or inhibition by α subunits [8]. Critical to our understanding of the cyclase system is the knowledge that all of the cyclase molecules cloned thus far are 12-membrane spanning proteins constructed of two distinct domains (termed M1 and M2). Although the regulatory aspects of these domains have not been fully explored, I have postulated that the two domains have distinct and specific interactions with the α and $\beta\gamma$ complexes [9]. This postulate requires dissociation of the heterotrimeric complex into α and $\beta\gamma$ subunits. As suggested from earlier modeling of the glucagon-sensitive cyclase system in liver membranes [10], hydrolysis of GTP to GDP and Pi is essential for the dynamic regulation of the adenylyl cyclase system and is a Mg-requiring process associated only with the high activity state of the system (designated E''). To accommodate the transition-state model, Fig. 1B depicts three distinct states of adenylyl cyclase: (1) State E represents the enzyme when unassociated with the GTP-activated monomer released from the oligomeric structure of stimulatory G-proteins (Gs); (2) E', a transition state formed by association of the monomer; and (3) E'', the high activity state formed by GTP-induced dissociation of the bound monomer to its α and $\beta\gamma$ products, a Mg-dependent process; these units combine with the M1 and M2 domains of cyclase. In this model, the rate-limiting step in the decay of the enzyme to the quiescent state (E) is the release of bound Pi and re-association of the resultant GDP-bound α subunit with the $\beta\gamma$ complex. The resultant GDP-bound heterotrimer binds with preferential affinity to the end of the oligomeric G-protein chain opposite to that from which the GTP-bound heterotrimer is released. The entire hormone activation process thus involves two cycles acting independently within the confines of the surface membrane. Although complex compared to other models of hormonal activation, the two cycles account for several well-known properties of hormone-sensitive G-protein-coupled systems: (a) The ability of a single receptor to activate multiple G-proteins; (b) Distinctive regulation of cyclases by α and $\beta\gamma$ depending on the type; and (c) Dynamic regulation of adenylyl cyclase systems in response to hormones, GTP, and Mg ions which, as suggested from modeling the cyclase system [10], is likely to occur through the transition state depicted here. An additional outgrowth of this theory is an explanation for the regulation of G-protein-linked systems by aluminum- or Mg-fluoride complexes. Binding of the fluoride complexes at the Mg-phosphate binding site could explain their pseudo-irreversible effects which necessitate accompanying binding of GDP to its site on the α subunits [11].

Clearly, there are many facets of this model to be explored in detail. Fortunately, most of the tools are now available for study at the molecular level. Critical, however, are questions that can only be answered by investigations of the native membrane-bound forms of signal transduction systems involving G-proteins. What factors determine the formation and organization of oligomeric G-proteins? What are the topological relationships between oligomeric G-proteins and adenylyl cyclase systems as well as other effector systems, including ion channels, phosphodiesterases, kinases and phosphatases? Recent studies indicate that adenylyl cyclases present in various parts of rat brain are highly localized in pre- and post-synaptic vesicles (Cooper, personal communication). Possibly these structures represent caveolae, invaginations of the surface membrane of cells that appear to contain a number of regulatory processes such as protein kinases and phosphatases, ion channels, and G-proteins [12]. Incorporation of such regulatory processes within the confines of caveolae may explain the concentration of signals (cyclic AMP and calcium ions). When "pinched off" as endocytic vesicles, circulation of these signals to selective regions of the cell may explain site-specific regulation of metabolic processes [13].

I am sure that Albert Renold would have been an incisive and probing critic of the thesis posed here. As one who enjoyed communal relationships, he probably would have appreciated my thoughts concerning organization of hormone-sensitive systems. In many respects these thoughts reflect the societal or self-organizational nature of all living organisms which, transposed to the molecular design of life's communication systems, is represented by DNA, RNA, cytoskeletal filamental proteins, and oligomeric G-proteins. These highly organized, polymeric systems are the means by which the living process defies the second law of thermodynamics. Through the orderly processing of these polymeric structures, discrete and selective bits of information enable the living organism or cell to manifest its adaptation to the environment without the entropic losses observed in non-living processes.

References

1. Baenziger JE, Jarrell HC, Hill RJ, Smith IC (1991) Average structural and rotational properties of a diunsaturated acyl chain in a lipid bilayer: effects of two cis-unsaturated double bonds. Biochemistry 30: 894–903
2. Rodbell M (1992) The role of GTP-binding proteins in signal transduction: from the sublimely simple to the conceptually complex. Curr Top Cell Regul 32: 1–47
3. Rodbell M (1980) The role of hormone receptors and GTP-regulatory proteins in membrane transduction. Nature 284: 17–22
4. Coulter S, Rodbell M (1992) Heterotrimeric G proteins in synaptoneurosome membranes are crosslinked by p-phenylenedimaleimide yielding structures comparable in size to crosslinked tubulin and F-actin. Proc Natl Acad Sci USA 89: 5842–5846
5. Jahangeer S, Rodbell M (1993) The disaggregation theory of signal transduction revisited: Further evidence that G-proteins are multimeric and disaggregate to monomers when activated. Proc Natl Acad Sci USA 90: 8782–8786
6. Rodbell M, Krans HM, Pohl SL, Birnbaumer L (1971) The glucagon-sensitive adenyl cyclase system in plasma membranes of rat liver. IV. Effects of guanylnucleotides on binding of 125I-glucagon. J Biol Chem 246: 1872–1876
7. Berstein G, Blank JL, Jhon DK, Exton JH, Rhee SG, Ross EM (1992) Phospholipase C-β1 is a GTPase-activating protein for Gq/11, its physiologic regulator. Cell 70: 411–418
8. Devivo M, Iyengar R (1994) G protein pathways – signal processing by effectors. Mol Cell Endocrinol 100: 65–70
9. Rodbell M, Jahangeer S, Coulter S (1993) G-proteins have properties of multimeric proteins: an explanation for the role of GTPases in their dynamic behavior. In: Dickey BF, Birnbaumer L (eds) GTPases in biology II. Springer, Berlin Heidelberg New York, pp 3–14 (Handbook of experimental pharmacology vol 108/II)
10. Rendell MS, Rodbell M, Berman M (1977) Activation of hepatic adenylate cyclase by guanyl nucleotides. Modeling of the transient kinetics suggests an "excited" state of GTPase is a control component of the system. J Biol Chem 252: 7909–7912

11. Spiegel AM, Downs RW, Aurbach GD (1979) Separation of a guanine nucleotide regulatory unit from the adenylate cyclase complex with GTP-affinity chromatography. J Cyclic Nucleotide Res 5: 3–17
12. Lisanti MP, Tang ZL, Scherer PE, Kubler E, Koleske AJ, Sargiacomo M (1995) Caveolae, transmembrane signalling and cellular transformation. Mol Membr Biol 12: 121–124
13. Haraguchi K, Rodbell M (1991) Carbachol-activated muscarinic (M1 and M3) receptors transfected into Chinese hamster ovary cells inhibit trafficking of endosomes. Proc Natl Acad Sci USA 88: 5964–5968

Anabolic Response to Cell Swelling in the Liver

L. Hue, V. Gaussin, and U. Krause

"At best, we are slowly reaching the stage at which we are aware of how little we know".
David Weatherall,
in *Science and the Quiet Art, 1995*

The study of the mechanisms responsible for the stimulation of liver glycogen synthesis by some amino acids has allowed us to discover that cell swelling was mediating not only a stimulation of glycogen synthesis but also lipogenesis. This led to the general idea that cell swelling triggers an anabolic response of the cell. This proposal was confirmed by the observation of its effect on protein synthesis/breakdown. This chapter briefly describes our wanderings through this field and summarizes our findings and the observations made by others. It also deals with the more recent concern with the anabolic signalling triggered by cell swelling.

Stimulation of Liver Glycogen Synthesis and Lipogenesis by Amino Acids

The liver can store relatively large amounts of glycogen, which is broken down and supplies blood with glucose during fasting periods. The rate of glycogen synthesis depends on the availability of substrates, and on the hormonal and dietary status of the organism. In isolated hepatocytes, glucose is a poor substrate for glycogen synthesis when given in the physiological range of concentrations, i.e., at $5-15\,\mathrm{m}M$ [1–3]. In these preparations, the rate of glycogen synthesis is at best 10% of the rate in vivo (about $1\,\mu$mol of glucose per minute and per gram of wet tissue), and the addition of insulin has proven to be of little help [4]. The synthesis is greatly enhanced by several gluconeogenic precursors, such as lactate, pyruvate or glycerol, and by several amino acids, such as glutamine, alanine, asparagine and proline [5]. With these precursors added together with glutamine, the rate of glycogen synthesis reaches physiological values. Besides their effect on glycogen synthesis, glutamine, proline and, to a lesser extent, alanine were also found to stimulate lipogenesis and to inhibit ketogenesis in isolated hepatocytes [6,7].

Several detailed studies have been performed to elucidate the mechanism whereby these amino acids stimulate glycogen synthesis. The following results were obtained: The amino acids do not act by providing carbon atoms for glycogen synthesis via gluconeogenesis, because 3-mercaptopicolinate, a known inhibitor of phosphoenolpyruvate carboxykinase (PEPCK), did not inhibit glycogen synthesis; on the contrary, it even stimulated the process [8]. Thus, it was proposed that stimulation of glycogen synthesis resulted from the build-up (or disappearance) of compounds whose concentration could be modified by these amino acids. The following effectors were proposed: the added amino acids themselves [6,9]; some unknown catabolites of glutamine, alanine or asparagine [5]; carbamoyl phosphate [10–12]; putative purine derivatives [6,13]; AMP [14]; and amino sugars [6]. A significant breakthrough in the understanding of the mechanism was achieved when a stimulation of glycogen synthesis was found with aminoisobutyric acid, a non-metabolizable amino acid analogue which is transported in a Na^+-dependent manner, like glutamine [9]. This led to the proposal that the Na^+-dependent transport and entry of amino acids, and/or the resulting ionic modifications were triggering glycogen synthesis. A well-known consequence of the Na^+-dependent entry of amino acids is the osmotic swelling of the cell [15,16]. Indeed, the transmembrane electrochemical gradient of Na^+ ions leads to the rapid intracellular accumulation of amino acids and of their metabolites, such as glutamate, which are poorly diffusible through the plasma membrane. Water enters the cells as a result of the overall intracellular accumulation of osmotically active substances and, hence, the volume of the cell increases. The experimental evidence demonstrating that the stimulation of glycogen synthesis and lipogenesis by amino acids is indeed mediated by cell swelling is summarized in the next section.

Activation of Glycogen Synthase and Acetyl-Coenzyme A Carboxylase by Cell Swelling

Glycogen synthesis and lipogenesis are controlled by the activity of glycogen synthase and acetyl-coenzyme A (CoA) carboxylase, respectively. Both enzymes are interconvertible by phosphorylation/dephosphorylation and are active in the dephosphorylated form. They possess multiple phosphorylation sites located at the N- and C-terminal domains. Phosphorylation of glycogen synthase on these sites follows a hierarchic order, with the phosphorylation of the primary sites preceding and allowing for the phosphorylation of the secondary sites, leading eventually to the inactivation of glycogen synthase [17]. On the other hand, acetyl-CoA carboxylase can be phosphorylated in vitro on as many as eight serine residues by six different protein kinases. However, in vivo, it is the rapid phosphorylation of Ser-79 by the AMP-activated protein kinase (AMP-PK), which differs from the cyclic AMP-dependent protein kinase, that inactivates acetyl-CoA carboxylase [18–20].

A single and direct relationship could be drawn between cell volume and glycogen synthesis (or glycogen synthase in the active form) when hepatocytes were incubated with those amino acids that are known to be co-transported with Na^+ ions [21]. Furthermore, and in agreement with this relationship, prevention of the glutamine-induced swelling by incubation of hepatocytes in hyperosmotic media hindered the activation of glycogen synthase. In addition, cell swelling induced by hypoosmotic media, even in the absence of amino acids, could also activate glycogen synthase; restoration of the isotonicity in such experiments suppressed both cell swelling and synthase activation [21].

Similar results were obtained with acetyl-CoA carboxylase, demonstrating that cell volume was also involved in the activation of this enzyme [22]. The striking similarities between the time-course of activation of glycogen synthase and acetyl-CoA carboxylase after stimulation by glutamine or proline [22] suggested the involvement of a common regulatory mechanism triggered by cell swelling. The current state of knowledge regarding this mechanism is described in the last section of this chapter.

Anabolic Response to Cell Swelling

Besides our work on liver glycogen synthesis and lipogenesis, several groups of workers and, in particular D. Häussinger and F. Lang, have performed a thorough study of the effects of cell volume on hepatic metabolism (for reviews, see [23–25]). A non-exhaustive list of the various effects of cell swelling is given in Table 1. The general picture that emerges from these studies is that cell swelling triggers an anabolic and anti-catabolic response of the cell. It stimulates glycogen synthesis [21,26,27], lipogenesis [22] and protein synthesis [23,28], whereas it inhibits glycogenolysis [23], ketogenesis [7,28,29] and proteolysis [23,30]. In addition, it stimulates amino acid transport and metabolism [31,32], urea synthesis [23], and bile acid excretion [31,33], and it increases the mRNA levels of ornithine decarboxylase [34], c-jun [35], actin [36] and tubulin [37], but it decreases PEPCK mRNA [38].

In the liver, the most common and frequent condition causing cell swelling is the postprandial phase of digestion. It corresponds to the cumulative uptake of amino acids that are co-transported with Na^+ and can be mimicked in livers perfused with a physiological mixture of amino acids [39]. These short-term changes in cell volume prime the liver for anabolic purposes.

Table 1. Effects of cell swelling on liver

Metabolism
Stimulates
    Glycogen synthesis
    Lipogenesis
    Protein synthesis
    Ureagenesis
    Bile acid excretion
·Inhibits
    Glycogenolysis
    Proteolysis
    Ketogenesis

2nd Messengers
Increase cAMP, Ca^{2+}, IP_3

Cytoskeleton
Stimulates the polymerization of actin
Stabilizes the microtubule network

Transport
Stimulates the transport of amino acids

Polyamines
Stimulate synthesis

cAMP, cyclic adenosine monophosphate; IP_3, inositol 1,4,5-triphosphate.

One may wonder whether cell swelling necessarily leads to an anabolic response of the cell. This is indeed not always the case and the nature of the determining agent seems crucial. Cell swelling may reflect some pathological conditions (lowering of temperature, anoxia) in which catabolism prevails. Moreover, the anabolic effects induced by glutamine are not entirely mimicked by hypoosmotic media. Indeed, the stimulation of lipogenesis by glutamine is greater than that found with hypoosmotic media, although the extent of activation of acetyl-CoA carboxylase was the same in both conditions [22]. This suggests that the metabolic situation induced by glutamine is relatively unique.

Glutamine: A Special Case

The metabolism of glutamine is characterized by its diversity. Glutamine is not only a constituent of proteins but it also participates in various metabolic pathways, either as a source of glutamate or as a donor of nitrogen. Such reactions include synthesis of purine and pyrimidine rings, formation of hexosamines, nicotinamide adenine dinucleotide (NAD^+) and asparagine. Glutamine also participates in transamination reactions [40,41]. In addition to its role as a precursor in metabolic pathways, glutamine serves as a fuel for fast-growing cells e.g., epithelial cells of the gut, stimulated lymphocytes, cultured fibroblasts and malignant cells [42–44]. Glutamine is not only a preferred respiratory fuel for such cells, but also a source of nitrogen for their biosynthetic processes. In such cells, glutamine is an essential substrate for growth and is routinely included in culture media [44].

Glutamine also plays a role in the detoxification of ammonia and in pH homeostasis. It is involved in the transport of carbon atoms and ammonia between extrahepatic tissues and the viscera. Moreover, the hepatic periportal–perivenous intercellular glutamine cycle [45,46] provides an effective means for the complete removal of portal ammonia. This mechanism permits adaptive changes in the rates of urea and glutamine production, and in the rate of bicarbonate removal by urea synthesis.

Glutamine is more than a simple amino acid metabolized in various pathways. It is endowed with anabolic properties which favour net accumulation of lipid, carbohydrate and protein. In short, glutamine bears an anabolic signal and, as such, is related to hormones. In the classical sense, hormones and neurotransmitters are chemical messengers secreted in trace amounts with functions restricted to control. Obviously glutamine does not fulfill these criteria (although two metabolically related molecules, the neurotransmitters glutamate and γ-aminobutyric acid, do), because it possesses other functions linked to its metabolism proper. Nevertheless, one may wonder whether glutamine does not belong to a new class of hormone-like molecules whose uptake triggers an anabolic response.

Mediation of Glucagon and Insulin Action by Changes in Cell Volume

Cell swelling may also reflect hormone action. Indeed, the anabolic response induced by cell swelling presents some striking similarity with the anabolic effects of insulin, the more so since insulin induces cell swelling [23,26,27]. Similarly, glucagon was reported to cause cell shrinkage and is known to induce a catabolic response that is mimicked by cell shrinkage resulting from hyperosmotic conditions. This applies for glycogen and protein metabolism (for reviews see [23–25]). The obvious question is therefore to know whether cell volume changes are the cause or the consequence of hormonal action.

It is generally accepted that the second messenger of glucagon action on carbohydrate metabolism is cyclic AMP. This messenger activates the cyclic AMP-dependent protein kinase, which, in turn, activates (or inactivates) key enzymes by phosphorylation. A study of the time-course of these changes following glucagon treatment of isolated hepatocytes showed that the increase in cyclic AMP content and glycogen phosphorylase activation were independent from volume changes and that, when it occurred, cell shrinkage followed glycogen breakdown [47]. Under these conditions, cell shrinkage is an epiphenomenon and is not required for the short-term effect of glucagon. The same conclusion could be drawn for the short-term effect of insulin on hepatocytes in suspension, i.e., inhibition of the effect of suboptimal doses of glucagon on carbohydrate metabolism [47].

The experiments reported above deal with the short-term effects of glucagon and insulin on carbohydrate metabolism. However, for the long-term effects of glucagon or insulin on the metabolism of both carbohydrate and protein, we can not exclude the possibility that cell swelling could participate in the hormonal action.

Transduction of the Anabolic Signal

Regulatory Volume Decrease

The experimental evidence so far obtained indicates that hepatocyte swelling, be it induced by certain amino acids or by incubation in hypoosmotic media, activates both glycogen synthase and acetyl-CoA carboxylase. We suggested, as a hypothesis, that the mechanism linking swelling to the activation of both enzymes is part of the cellular response to the osmotic stress. This cellular response aims at restoring the initial cell volume and it has been studied in detail in several cell types [48]. Hepatocytes incubated under hypoosmotic conditions first swell and then undergo an intricate mechanism of regulatory volume decrease, which eventually leads to an electrogenic K^+ efflux. This efflux results from the activation of K^+ channels, itself caused by an increase in Ca^{2+} concentration or by the stretch imposed upon the plasma membrane by swelling. Consequently, Cl^- ions, which are permeant ions in hepatocytes, distribute across the plasma membrane according to the membrane potential. A fall in intracellular concentration of Cl^- ions is thus observed in swollen hepatocytes.

Incubation of hepatocytes with glutamine or other Na^+-cotransported amino acids leads to additional changes in the ionic composition of the cells. Addition of glutamine not only induces the drop in intracellular Cl^- ions, but it also increases the intracellular concentration of glutamate.

Variations in the concentrations of ions such as K^+, Ca^{2+}, Cl^- and glutamate could lead to an activation of glycogen synthase and acetyl-CoA carboxylase following cell swelling. Similarly, changes in the concentration of cyclic AMP or in the intracellular pH could be involved in the activation process.

K⁺ Ions and Intracellular pH

A mechanism involving K^+ ions is unlikely because K^+ ions and amino acids seem to act differently, despite their similar effects on cell swelling, glycogen synthesis and glycogen synthase activation. Indeed, K^+ ions, but not amino acids, promote glycogen phosphorylase inactivation [2] and stimulate glucose phosphorylation [49]. Moreover, amino acids and hypoosmotic media tend to decrease K^+ concentration [48,50].

A change in intracellular pH is another possible mechanism and intracellular pH was found to be slightly decreased (from 7.26 to 7.19) under hypoosmotic conditions [51]. Incubation of cultured hepatocytes at pH 7.1 instead of 7.4 slightly stimulated glycogen synthesis [27]. Our results did not confirm this observation and modulation of the extracellular pH between 7.2 and 7.6 had no detectable effect on glycogen synthesis in isolated hepatocytes (A. Baquet, L. Garcia-Salguero and L. Hue, unpublished work).

Cyclic AMP and Ca²⁺ Ions

A decrease in free cytosolic Ca^{2+} concentration might be responsible for an activation of glycogen synthase. It should also inactivate glycogen phosphorylase. This, however, has not been observed with glutamine [6,52]. Actually, an increase rather than a decrease in Ca^{2+} concentration was found in cells submitted to a hypoosmotic shock [53,54], and in hepatocytes incubated with glutamine [52]. An increase in the concentration of $Ins(1,4,5)P_3$ and cAMP was also observed [52,55], yet the activation state of phosphorylase was not affected. Moreover, the activation of glycogen phosphorylase by suboptimal concentrations of vasopressin or angiotensin II was even partly antagonized by amino acids or hypoosmolarity, despite the basal increase in Ca^{2+} and cAMP concentrations [52]. This suggests that cell swelling induces changes in the concentration of effectors that are able to overcome, at least in part, the effect of Ca^{2+} ions and cAMP on glycogen phosphorylase activation. These effectors could be either Cl^- ions or metabolites derived from amino acids such as glutamate.

Cl⁻ Ions and Glutamate

The concentration of Cl^- ions falls from 35–40 mM in control hepatocytes down to 10–15 mM in swollen hepatocytes [56]. In hepatocytes incubated with 10 mM glutamine, the final concentration of glutamate can be as high as 20 mM [56,57]. Therefore, the effect of Cl^- ions and glutamate on the activation of glycogen synthase and acetyl-CoA carboxylase, was studied in vitro with liver extracts [56,58]. In this system, high concentrations (above 50 mM) of KCl completely inhibited glycogen synthase activation, and inhibited by about 20% acetyl-CoA carboxylase activation, whereas low concentrations (below 15 mM) of KCl had no detectable effect. Glutamate stimulated in a dose-dependent manner the activation of both enzymes. With 20–25 mM glutamate, a concentration found in hepatocytes incubated with glutamine, the stimulation was more than 500% for acetyl-CoA carboxylase, but only 30%–50% for glycogen synthase.

Glutamate-Activated Type 2A Protein Phosphatase

The activation of glycogen synthase and acetyl-CoA carboxylase by cell swelling could be mediated either by a stimulation of protein phosphatase(s), an inhibition of protein kinase(s), or a combination of both. The involvement of protein phosphatases is demonstrated by the inhibition of the activation of both enzymes by microcystin, a

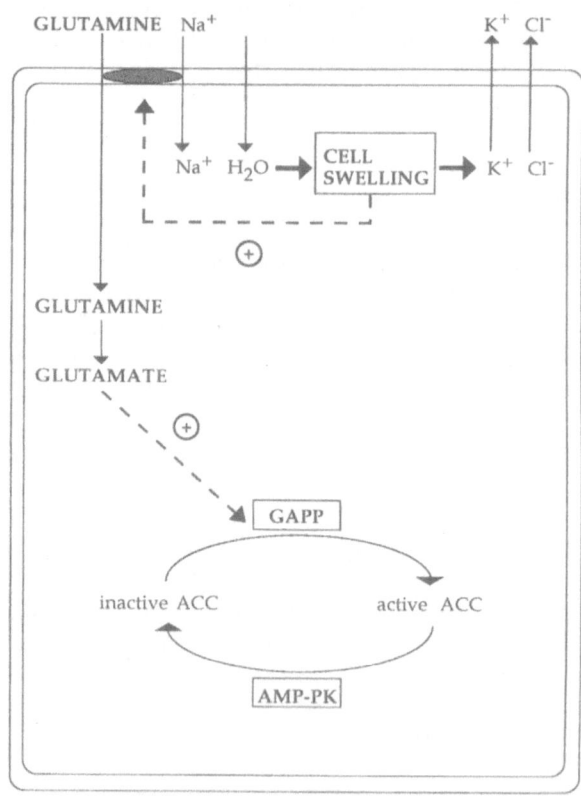

Fig. 1. Proposed mechanism of activation of acetyl-coenzyme A (CoA) carboxylase (*ACC*) by glutamine-induced cell swelling. The Na⁺-dependent entry of glutamine into the cell leads to the intracellular accumulation of osmotically active substances and to cell swelling. The latter further stimulates glutamine transport in a positive feedback. The activation of acetyl-CoA carboxylase results from the stimulation of the glutamate-activated type 2A protein phosphatase (*GAPP*), which activates the enzyme by dephosphorylation. It is proposed that GAPP is antagonizing AMP-activated protein kinase (*AMP-PK*) which inactivates ACC by phosphorylation. Glycogen synthase could be activated by a similar mechanism (see text for further explanation)

known inhibitor of protein phosphatases [58]. Although this does not rule out some protein kinase-dependent mechanism, our first approach was to identify the protein phosphatase acting on acetyl-CoA carboxylase.

This protein phosphatase has been purified thanks to its sensitivity towards glutamate [59]. It is a type 2A protein phosphatase according to P. Cohen's classification [60]. The activity of the purified enzyme depends on the presence of dicarboxylic acids such as glutamate and is stimulated by Mg^{2+} ions. In addition, it was found to be able to dephosphorylate a synthetic peptide corresponding to a sequence of acetyl-CoA carboxylase that contains Ser-79 which is phosphorylated by AMP-PK. This protein kinase is known to phosphorylate and inactivate not only acetyl-CoA carboxylase but also glycogen synthase and hydroxymethylglutaryl-CoA (HMG-CoA) reductase, the key enzyme in cholesterol synthesis.

We therefore propose that the glutamate-activated type 2A protein phosphatase antagonizes the action of AMP-PK on acetyl-CoA carboxylase (Fig. 1). If this glutamate-activated protein phosphatase possesses the same substrate specificity as that of AMP-PK, then a common enzymatic mechanism can be proposed for the activation of glycogen synthase and acetyl-CoA carboxylase. This mechanism, which underlines the crucial role played by glutamate, could also apply to the activation of HMG-CoA reductase and, hence, to cholesterol synthesis.

Mitogen-Activated Protein Kinases and Osmotic Stress

The above considerations and findings do not necessarily rule out the involvement of a protein kinase cascade in the anabolic signalling by cell swelling. Indeed, in hepatocytes that are allowed to swell under hypoosmotic conditions, but in the absence of amino acids, glutamate does not accumulate and cannot be involved in the activation of acetyl-CoA carboxylase. In this case, the fall in intracellular Cl⁻ ions is probably insufficient to explain the activation observed in intact cells, since the stimulation of the activation of the enzyme resulting from the decrease in Cl⁻ ions is relatively small. Therefore, other mechanisms must be involved for the activation of acetyl-CoA carboxylase in hepatocytes incubated in hypoosmotic media.

Several lines of evidence obtained in various cell types, including hepatocytes, have validated this proposal and indicate that the mitogen-activated protein kinases (MAPK) could also be involved in the response of the cell to osmotic stress. A number of MAPKs have been identified. They represent a superfamily of serine/threonine protein kinases that are involved in various cellular processes in response to stimuli such as hormones, growth factors, tumour promoters and stress conditions (Fig. 2). Their substrates are numerous and include, among others, transcription factors (c-Fos, c-Jun), cytoskeletal proteins, downstream protein kinases (MAPK-activated protein kinases [MAPKAP-K1, MAPKAP-K2]), other proteins (myelin basic protein, a substrate that is commonly used for the in vitro assay of MAPK), enzymes (phospholipase A2) and receptors (e.g. the epidermal growth factor (EGF) receptor) (for a review see [61]).

An activation of MAPK – or more precisely, myelin basic protein kinase – has been reported in rat hepatocyte monolayers that were made to swell either by incubation in hypoosmotic media or with proline [62]. More recently, it has been found that cell

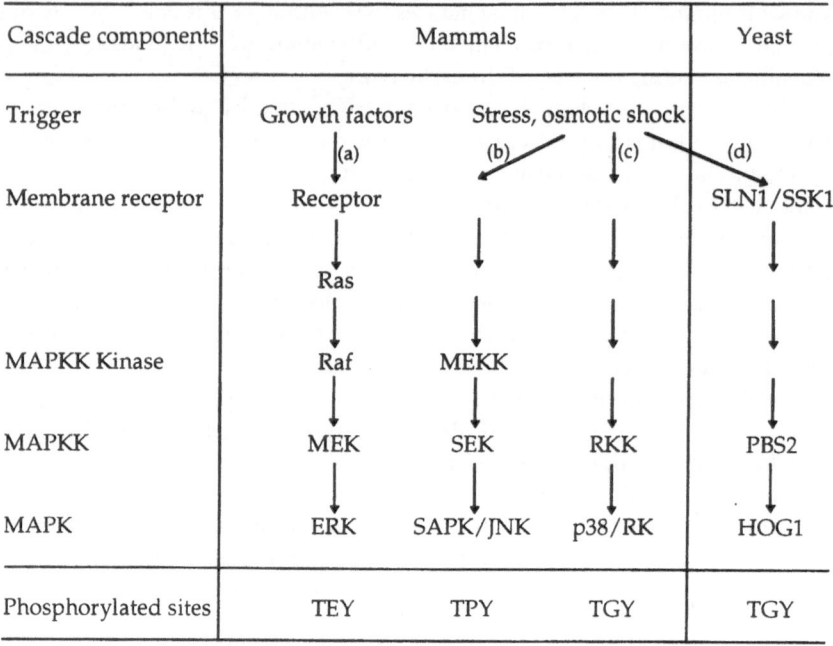

Cascade components	Mammals				Yeast
Trigger	Growth factors	Stress, osmotic shock			
	(a)	(b)	(c)		(d)
Membrane receptor	Receptor				SLN1/SSK1
	Ras				
MAPKK Kinase	Raf	MEKK			
MAPKK	MEK	SEK	RKK		PBS2
MAPK	ERK	SAPK/JNK	p38/RK		HOG1
Phosphorylated sites	TEY	TPY	TGY		TGY

Fig. 2. Mitogen-activated protein kinase (*MAPK*) cascades involved in growth factor action and in the cellular responses to stress, including osmotic shock, in mammals and yeast. The MAPK cascades involve at least three different levels of protein kinases. These include MAPK-kinase-kinase (*MAPKK-kinase*) which activates by phosphorylation MAPK kinase (MAPKK) which is a dual specificity protein kinase phosphorylating MAPK on threonine and tyrosine residues. The sequences of the phosphorylated sites differ and are indicated in the *last line* of the scheme. The MAPK cascade involved in growth factor action is shown in (*a*). The cascades involved in cellular response to stress and osmotic shock are given in (*b*) and (*c*). The related cascade occurring in yeast is depicted in (*d*). Abbreviations of the protein kinases: in (*a*), *ERK*, extracellular signal regulated kinase; *MEK*, MAPK and ERK kinase; in (*b*), *SAPK*, stress-activated protein kinase; *JNK*, c-Jun N-terminal kinase; *SEK*, SAPK and ERK kinase; *MEKK*, MEK kinase; in (*c*), *RK*, reactivation kinase; *RKK*, *RK* kinase; in (*d*), HOG1, high osmolarity glycerol response kinase 1

swelling induced by amino acids, known to inhibit proteolysis, increases the phosphorylation state of ribosomal S6 protein in isolated hepatocytes [63]. One of the S6 protein kinases, MAPKAP-K1 or p90S6K, is known to be activated by MAPK. Moreover, hypoosmotic exposure of hepatoma cells changes the electrophoretic mobility of the MAPKs Erk-1 and Erk-2 (extracellular signal regulated kinases). This has been interpreted as a change in phosphorylation and activation states of these MAPKs [25]. Our recent results [64] do not confirm such an activation of Erks in hepatocytes incubated with glutamine or in hypoosmotic media. They, however, indicate that hepatocyte swelling activates phosphatidylinositol 3-kinase (PI3 kinase), an enzyme involved in the insulin signalling pathway (Fig. 3).

In this context, it is thus worth noting that the activation of muscle glycogen synthase by insulin is proposed to result from a stimulation of the MAPK cascade (Fig. 3). MAPK phosphorylates and activates MAPKAP-K1, which in turn phosphorylates and inactivates glycogen synthase kinase 3 (GSK-3, one of the glycogen synthase

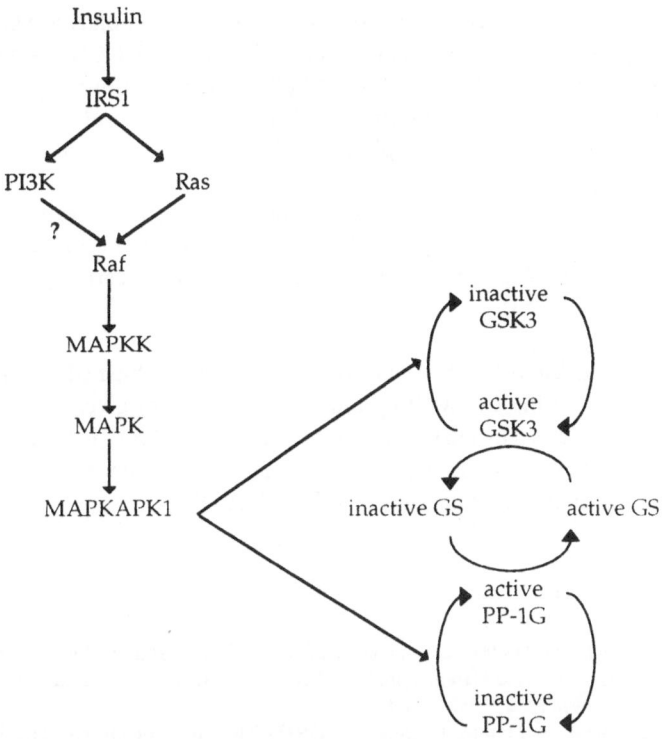

Fig. 3. Proposed mechanism of activation of glycogen synthase by insulin in the muscle. The mechanism involves a mitogen-activated protein kinase (*MAPK*) cascade ending in the activation of MAPK activated protein kinase 1 (*MAPKAPK1*) which inactivates glycogen synthase kinase 3 (*GSK3*) and activates type 1 G-protein phosphatase (*PP-1G*) by phosphorylation. The activation of the MAPK cascade seems to be mediated by Ras and/or by phosphatidylinositol 3 kinase (*PI₃K*). *IRS₁*, insulin receptor substrate 1

inactivating enzymes) and activates type-1G protein phosphatase (PP1G, a glycogen synthase activating enzyme) [65–67]. The end result of this phosphorylation cascade is the activation of glycogen synthase. A similar cascade is currently being considered for the activation of glycogen synthase by cell swelling in the liver [64].

Other experimental evidence suggests that MAPKs could be involved in the cellular response to osmotic stress. In the human intestine 407 cell line, tyrosine phosphorylation is an essential step in the regulatory volume decrease response. In these cells, hypoosmotic shock triggered a rapid and transient increase in tyrosine phosphorylation of several proteins as well as phosphorylation of MAPK [68]. In cells of monocytic origin, a protein kinase, p38, which was tyrosine phosphorylated in response to endotoxic lipopolysaccharide, has been identified [69]. This protein presents 52% identity of amino acid sequence with that of a product of the *Saccharomyces cerevisiae* osmosensing gene HOG1 (high osmolarity glycerol response) and 46%–49% identity with members of the MAPK family [70,71]. Indeed, in yeast, the hybrid histidine kinase (SLN1) and the response regulator (SSK1) control the activation of a protein kinase cascade related to the MAPK cascade of the mammalian system, initiated by an increase in osmolarity. This leads to a stimulation of glycerol synthesis. The yeast cascade involves HOG1, a MAPK, and PBS2, a MAPK kinase (Fig. 2). Finally, other MAPKs [c-Jun N-terminal protein kinases (JNK) and stress-activated protein kinases (SAPK)] have been identified in mammalian tissues and were found to be phosphorylated in stress conditions or when osmolarity increases [72–75].

Mammalian tissues, therefore, possess several MAPKs that could participate in the anabolic response to cell swelling. A precise mechanism linking MAPKs to the activation of glycogen synthase and acetyl-CoA carboxylase remains, however, to be established.

Note Added in Proof. According to our recent published data [64], MAPK is probably not mediating the swelling-induced activation of glycogen synthase and acetyl-CoA carboxylase. The current hypothesis is that protein kinase B, the new protein kinase, also called RAC, which is downstream of PI 3 Kinase, participates in the transduction of the anabolic signal triggered by cell swelling.

References

1. Seglen PO (1974) Autoregulation of glycolysis, respiration, gluconeogenesis and glycogen synthesis in isolated parenchymal rat liver cells under aerobic and anaerobic conditions. Biochim Biophys Acta 338: 317–336
2. Hue L, Bontemps F, Hers HG (1975) The effect of glucose and of potassium ions on the interconversion of the two forms of glycogen phosphorylase and of glycogen synthetase in isolated rat liver preparations. Biochem J 152: 105–114
3. Katz J, Wals PA, Golden S, Rognstad R (1975) Recycling of glucose by rat hepatocytes. Eur J Biochem 60: 91–101
4. Stalmans W, van de Werve G (1981) Regulation of glycogen metabolism by insulin. In: Hue L, van de Werve G (eds) Short-term regulation of liver metabolism, Elsevier, Amsterdam, pp 119–138
5. Katz J, Golden S, Wals PA (1976) Stimulation of hepatic glycogen synthesis by amino acids. Proc Natl Acad Sci USA 73: 3433–3437
6. Lavoinne A, Baquet A, Hue L (1987) Stimulation of glycogen synthesis and lipogenesis by glutamine in isolated rat hepatocytes. Biochem J 248: 429–437
7. Baquet A, Lavoinne A, Hue L (1991) Comparison of the effects of various amino acids on glycogen synthesis, lipogenesis and ketogenesis in isolated rat hepatocytes. Biochem J 273: 57–62
8. Okajima F, Katz J (1979) Effects of mercaptopicolinic acid and of transaminase inhibitors on glycogen synthesis by rat hepatocytes. Biochem Biophys Res Commun 87: 155–162
9. Rognstad R (1986) Effects of amino acid analogs and amino acid mixtures on glycogen synthesis in rat hepatocytes. Biochem Arch 2: 185–190
10. Bode AM, Nordlie RC (1993) Reciprocal effects of proline and glutamine on glycogenesis from glucose and ureagenesis in isolated, perfused rat liver. J Biol Chem 268: 16298–16301
11. Rognstad R (1985) Possible role for carbamyl phosphate in the control of liver glycogen synthesis. Biochem Biophys Res Commun 130: 229–233
12. Gustafson L, Romp N, van Woerkom GM, Meijer AJ (1994) Carbamoyl phosphate and ureagenesis are not involved in amino-acid-stimulated glycogenesis. Eur J Biochem 223: 553–556
13. Solanki K, Moser U, Nyfeler F, Walter P (1982) Possible correlation between the stimulation of glycogen synthesis by some amino acids and the synthesis of purines in hepatocytes from starved rats. Experientia 38: 732
14. Carabaza A, Ricart MD, Mor A, Guinovart JJ, Ciudad CJ (1990) Role of AMP on the activation of glycogen synthase and phosphorylase by adenosine, fructose and glutamine in rat hepatocytes. J Biol Chem 265: 2724–2732

15. Bakker-Grunwald T (1983) Potassium permeability and volume control in isolated rat hepatocytes. Biochim Biophys Acta 731: 239–242
16. Kristensen LØ, Folke M (1984) Volume-regulatory K$^+$ efflux during concentrative uptake of alanine in isolated rat hepatocytes. Biochem J 221: 265–268
17. Roach PJ (1990) Control of glycogen synthase by hierarchal protein phosphorylation. FASEB J 4: 2961–2968
18. Hardie DG, Carling D, Sim ATR (1989) The ATP-activated protein kinase: a multisubstrate regulator of lipid metabolism. Trends Biochem Sci 14: 20–23
19. Davies SP, Sim ATR, Hardie DG (1990) Location and function of three sites phosphorylated on rat acetyl-CoA carboxylase by the AMP-activated protein kinase. Eur J Biochem 187: 183–190
20. Hardie DG (1992) Regulation of fatty acid and cholesterol metabolism by the AMP-activated protein kinase. Biochim Biophys Acta 1123: 231–238
21. Baquet A, Hue L, Meijer AJ, van Woerkom GM, Plomp PJAM (1990) Swelling of rat hepatocytes stimulates glycogen synthesis. J Biol Chem 265: 955–959
22. Baquet A, Maisin L, Hue L (1991) Swelling of rat hepatocytes activates acetyl-CoA carboxylase in parallel to glycogen synthase. Biochem J 278: 887–890
23. Häussinger D, Lang F (1991) Cell volume in the regulation of hepatic function: a mechanism for metabolic control. Biochim Biophys Acta 1071: 331–350
24. Häussinger D (1995) Regulation of metabolism by changes in cellular hydration. Clin Nut 14: 4–12
25. Häussinger D, Schliess F (1995) Cell volume and hepatocellular function. J Hepatol 22: 94–100
26. Al-Habori M, Peak M, Thomas TH, Agius L (1992) The role of cell swelling in the stimulation of glycogen synthesis by insulin. Biochem J 282: 789–796
27. Peak M, Al-Habori M, Agius L (1992) Regulation of glycogen synthesis and glycolysis by insulin, pH and cell volume. Interactions between swelling and alkalinization in mediating the effects of insulin. Biochem J 282: 797–805
28. Stoll B, Gerok W, Lang F, Häussinger D (1992) Liver cell volume and protein synthesis. Biochem J 287: 217–222
29. Guzman M, Velasco G, Castro J, Zammit VA (1994) Inhibition of carnitine palmitoyltransferase I by hepatocyte swelling. FEBS Lett 344: 239–241
30. Meijer AJ, Gustafson LA, Luiken JJ, Blommaart PJ, Caro LH, van Woerkom GM, Spronk C, Boon L (1993) Cell swelling and the sensitivity of autophagic proteolysis to inhibition by amino acids in isolated rat hepatocytes. Eur J Biochem 215: 449–454
31. Häussinger D, Hallbrucker C, Saha N, Lang F, Gerok W (1992) Cell volume and bile acid excretion. Biochem J 288: 681–689
32. Bode AM, Kilberg MS (1991) Amino acid-dependent increase in hepatic system N activity is linked to cell swelling. J Biol Chem 266: 7376–7381
33. Hallbrucker C, Lang F, Gerok W, Häussinger D (1992) Cell swelling increases bile flow and taurocholate excretion into bile in isolated perfused rat liver. Biochem J 281: 593–595
34. Tohyama Y, Kameji T, Hayashi S (1991) Mechanisms of dramatic fluctuations of ornithine decarboxylase activity upon tonicity changes in primary cultured rat hepatocytes. Eur J Biochem 202: 1327–1331
35. Finkenzeller G, Newsome WP, Lang F, Häussinger D (1994) Increase of c-jun mRNA upon hypoosmotic cell swelling of rat hepatoma cells. FEBS Lett 340: 163–166
36. Theodoropoulos PA, Stournaras C, Stoll B, Markogiannakis E, Lang F, Gravanis A, Häussinger D (1992) Hepatocyte swelling leads to rapid decrease of the G-/total actin ratio and increases actin mRNA levels. FEBS Lett 311: 241–245
37. Häussinger D, Stoll B, vom Dahl S (1994) Effect of hepatocyte swelling on microtube stability and tubulin mRNA levels. Biochem Cell Biol 72: 12–19
38. Newsome WP, Warskulat U, Noe B, Wettstein M, Stoll B, Gerok W, Häussinger D (1994) Modulation of phosphoenolpyruvate carboxykinase mRNA levels by the hepatocellular hydration state. Biochem J 304: 555–560
39. Wettstein M, vom Dahl S, Lang F, Gerok W, Häussinger D (1990) Cell volume regulatory responses of isolated perfused rat liver. Biol Chem Hoppe Seyler 371: 493–501
40. Krebs HA (1980) Glutamine metabolism in the animal body. In: Mora J, Palacios R (eds) Glutamine: metabolism, enzymology and regulation. Academic, New York, pp 319–329

41. Meister A (1980) Catalytic mechanism of glutamine synthetase; overview of glutamine meta-
    bolism. In: Mora J, Palacios R (eds) Glutamine: metabolism, enzymology and regulation.
    Academic, New York, pp 1-40
42. Kovacevic Z, Morris HP (1972) The role of glutamine in the oxidative metabolism of malignant
    cells. Cancer Res 32: 326-333
43. Windmueller HG (1980) Enterohepatic aspects of glutamine metabolism. In: Mora J, Palacios R
    (eds) Glutamine: metabolism, enzymology and regulation. Academic, New York, pp 235-257
44. Zielke HR, Zielke CI, Ozand PT (1984) Glutamine: a major energy source for cultured mammalian
    cells. Fed Proc 43: 121-125
45. Häussinger D (1983) Hepatocyte heterogeneity in glutamine and ammonia metabolism and the
    role of an intercellular glutamine cycle during ureagenesis in rat liver. Eur J Biochem 133: 269-275
46. Häussinger D (1986) Regulation of hepatic ammonia metabolism: the intercellular glutamine
    cycle. Adv Enzyme Regul 25: 159-180
47. Gaussin V, Baquet A, Hue L (1992) Cell shrinkage follows rather than mediates the short-term
    effects of glucagon on carbohydrate metabolism. Biochem J 287: 17-20
48. Hoffmann EK, Simonsen LO (1989) Membrane mechanisms in volume and pH regulation in
    vertebrate cells. Physiol Rev 69: 315-382
49. Bontemps F, Hue L, Hers HG (1978) Phosphorylation of glucose in isolated rat hepatocytes.
    Sigmoidal kinetics explained by the activity of glucokinase alone. Biochem J 174: 603-611
50. Lang F, Stehle T, Häussinger D (1989) Water, K^+, H^+, lactate and glucose fluxes during cell volume
    regulation in perfused rat liver. Pflugers Arch 413: 209-216
51. Gleeson D, Corasanti JG, Boyer JL (1990) Effects of osmotic stresses on isolated rat hepatocytes.
    II. Modulation of intracellular pH. Am J Physiol 258: G299-G307
52. Baquet A, Meijer AJ, Hue L (1991) Hepatocytes swelling increases inositol 1,4,5-trisphosphate,
    calcium and cyclic AMP concentration but antagonizes phosphorylase activation by Ca^{3+}-depen-
    dent hormones. FEBS Lett 278: 103-106
53. Watson PA (1989) Accumulation of cAMP and calcium in S49 mouse lymphoma cells following
    hypoosmotic swelling. J Biol Chem 264: 14735-14740
54. Hazama A, Okada Y (1990) Involvement of Ca^{2+}-induced Ca^{2+} release in the volume regulation
    of human epithelial cells exposed to a hypotonic medium. Biochem Biophys Res Commun 167:
    287-293
55. vom Dahl S, Hallbrucker C, Lang F, Häussinger D (1991) Role of eicosanoids, inositol phosphates
    and extracellular Ca^{2+} in cell volume regulation of rat liver. Eur J Biochem 198: 73-83
56. Meijer AJ, Baquet A, Gustafson L, van Woerkom G, Hue L (1992) Mechanism of activation of liver
    glycogen synthase by swelling. J Biol Chem 267: 5823-5828
57. Plomp PJAM, Boon L, Caro LHP, van Woerkom GM, Meijer AJ (1990) Stimulation of glycogen
    synthesis in hepatocytes by added amino acids is related to the total intracellular content of amino
    acids. Eur J Biochem 191: 237-243
58. Baquet A, Gaussin V, Bollen M, Stalmans W, Hue L (1993) Mechanism of activation of liver acetyl-
    CoA carboxylase by cell swelling. Eur J Biochem 217: 1083-1089
59. Gaussin V, Hue L, Stalmans W, Bollen M (1995) The activation of hepatic acetyl-CoA carboxylase
    by glutamate and Mg^{2+} is mediated by protein phosphatase-2A. Biochem J 316: 217-224
60. Cohen P (1989) The structure and regulation of protein phosphatases. Annu Rev Biochem
    58: 453-508
61. Denton RM, Tavaré JM (1995) Does mitogen-activated protein kinase have a role in insulin
    action? Eur J Biochem 227: 597-611
62. Agius L, Peak M, Beresford G, Al-Habori M, Thomas TH (1994) The role of ion content and cell
    volume in insulin action. Biochem Soc Trans 22: 516-522
63. Blommaart EFC, Luiken JJFP, Blommaart PJE, van Woerkom GM, Meijer AJ (1995) Phosphory-
    lation of ribosomal protein S6 is inhibitory for autophagy in isolated rat hepatocytes. J Biol
    Chem 270: 2320-2326
64. Krause K, Rider MH, Hue L (1996) Protein kinase signalling pathway triggered by cell swelling
    and involved in the activation of glycogen synthesis and acetyl-CoA carboxylase in isolated rat
    hepatocytes. J Biol Chem 271: 16668-16673
65. Dent P, Lavoinne A, Nakielny S, Caudwell FB, Watt P, Cohen P (1990) The molecular
    mechanism by which insulin stimulates glycogen synthesis in mammalian skeletal muscle. Nature
    348: 302-308

66. Lavoinne A, Erikson E, Maller JL, Price DL, Avruch J, Cohen P (1991) Purification and characterization of the insulin-stimulated protein kinase from rabbit skeletal muscle; close similarity to S6 kinase II. Eur J Biochem 199: 723–728

67. Cross DAE, Alessi DR, Vandenheede JR, McDowell HE, Hundal HS, Cohen P (1994) The inhibition of glycogen synthase kinase-3 by insulin or insulin-like growth factor 1 in the rat skeletal muscle cell line L6 is blocked by wortmannin, but not by rapamycin: evidence that wortmannin blocks activation of the mitogen-activated protein kinase pathway in L6 cells between Ras and Raf. Biochem J 303: 21–26

68. Tilly BC, van den Berghe N, Tertoolen LGJ, Edixhoven MJ, de Jonge HR (1993) Protein tyrosine phosphorylation is involved in osmoregulation of ionic conductances. J Biol Chem 268: 19919–19922

69. Han J, Lee J-D, Bibbs L, Ulevitch RJ (1994) A MAP kinase targeted by endotoxin and hyperosmolarity in mammalian cells. Science 265: 808–811

70. Brewster JL, de Valoir T, Dwyer ND, Winter E, Gustin MC (1993) An osmosensing signal transduction pathway in yeast. Science 259: 1760–1763

71. Maeda T, Wurgler-Murphy SM, Saito H (1994) A two-component system that regulates an osmosensing MAP kinase cascade in yeast. Nature 369: 242–245

72. Galcheva-Gargova Z, Dérijard B, Wu I-H, Davis RJ (1994) An osmosensing signal transduction pathway in mammalian cells. Science 265: 806–808

73. Rouse J, Cohen P, Trigon S, Morange M, Alonso-Llamzares A, Zamanillo D, Hunt T, Nebreda AR (1995) A novel kinase cascade triggered by stress and heat shock that stimulates MAPKAP kinase-2 and phosphorylation of the small heat shock proteins. Cell 78: 1027–1037

74. Swanson RV, Alex LA, Simon MI (1994) Histidine and aspartate phosphorylation: two-component systems and the limits of homology. TIBS 19: 485–490

75. Cano E, Mahadevan LC (1995) Parallel signal processing among mammalian MAPKs. Trends Biochem Sci 20: 117–122

Intercellular Communication and Insulin Secretion

P. Meda

Introduction

In 1973, a Geneva group reported for the first time that the endocrine cells of the islets of Langerhans are linked by gap junctions and suggested that these membrane structures may be somehow involved in the control of insulin secretion [1]. This work, which has set the basis for most of the later studies I review here, was senior authored by Prof. A.E. Renold. As on several other occasions, he, to whom I was later introduced as "Uncle Albert," had timely sensed with L. Orci the potential importance of this novel form of cell-to-cell interaction. Throughout the 11 years I had the chance to see Albert Renold in Geneva, his enthusiasm and warm support of my own work in the field of gap junctional communication never failed.

Proper functioning of multicellular systems necessarily depends on a communication network for cross-talk between cells. Via such a network, individual cells sense the state of activity of their neighbours, and accordingly regulate their own level of functioning. Vertebrate cells may communicate in a variety of ways, which differ in the type of structures and mechanisms used to establish cellular cross-talk, as well as in the nature of the signals exchanged by the communicating cells [2–5].

A widespread mechanism for intercellular communication is via the diffusion in the extracellular space of hormones and neuromediators that are simultaneously sensed by cells bearing cognate receptors (Fig. 1). This system ensures highly specific signalling between both distant (hormonal and neural communication) and nearby cells (paracrine communication). A second form of cell-to-cell communication is provided by the diffusion in the extracellular space of ions and molecules which enter cells either by free diffusion through the lipid membrane bilayer or by way of specific channels and transporters (Fig. 1). Coordination is then achieved by the simultaneous activation of specific metabolic and effector steps in several cells. In spite of the diversity of the structures and signals used to establish cellular cross-talk, all these communication mechanisms require the passage of signals through extracellular spaces. Since these spaces mediate the transfer of information from one cell to another, these forms of communication are referred to as "indirect". Most cell types also communicate in a "direct" way, i.e., by mechanisms which do not involve the passage of signals in the extracellular space, and which obligatorily require a contact between the interacting cells. One way of establishing direct cell-to-cell communication is via cell adhesion molecules [4], the surface glycoproteins which play a prominent role in establishing and maintaining cell contacts (Fig. 1). These molecules are well suited to transmit information since they closely approximate the membranes of adjacent cells and associate with cytoskeletal elements which act as information transducers at both cytoplasm and nucleus levels. Direct cell-to-cell communication is also mediated by connexins [3,6], the nonglycosylated proteins which form gap junctions (Fig. 1). These structures are clusters of highly permeable channels that cross the membranes of adjacent cells and bridge the intervening extracellular space, thus allowing for a diffusional cell-to-cell exchange of cytoplasmic ions and small molecules.

The secretory cells of pancreatic islets communicate in a variety of ways which presumably include all of the indirect and direct mechanisms summarized above. It has long been known that insulin-producing β-cells regulate their functions through interaction with hormones, neuromediators and other signals, which reach the pancreas from outside or which are generated within the gland itself [7–10]. The more recent recognition that insulin secretion is still a regulated event under in vitro conditions which perturb the native blood supply, innervation and flux of extracellular fluid indicates, however, that other communication mechanisms are also operative, if not predominant in β-cell control. The finding that insulin secretion is altered after dispersion of β-cells and rapidly improves after their reaggregation [11–18] further suggests that these mechanisms depend on the establishment and maintenance of proper cell-to-cell contacts.

Here, I shall briefly summarize the indirect communication which takes place between β-cells and which has already been extensively discussed in several recent

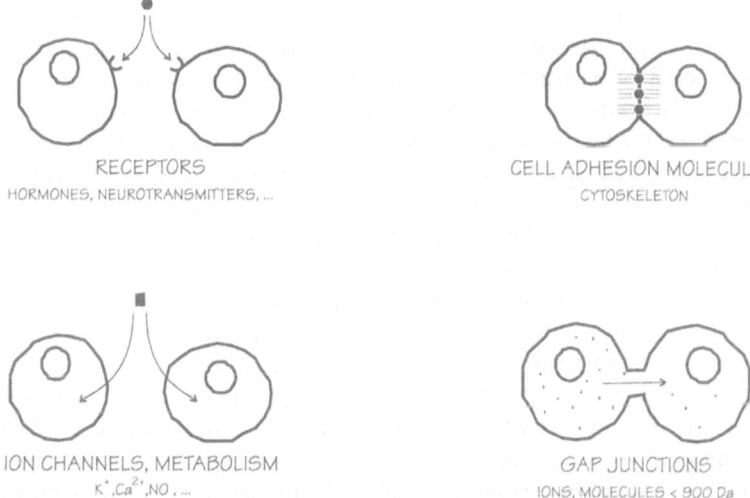

RECEPTORS
HORMONES, NEUROTRANSMITTERS, ...

CELL ADHESION MOLECULES
CYTOSKELETON

ION CHANNELS, METABOLISM
K^+, Ca^{2+}, NO, ...

GAP JUNCTIONS
IONS, MOLECULES < 900 Da

Fig. 1. Mechanisms for communication between β-cells. As with most other types of vertebrate cells, pancreatic β-cells may communicate by multiple mechanisms, which involve the exchange of signals transiting either in the extracellular spaces or between the cytoplasms of cells in close contact. *Left panel:* The first type of cell-to-cell communication, also referred to as indirect, includes all forms of hormonal and neural interactions, as well as the intercellular signalling mediated by current-carrying ions and molecules to which the cell membrane is permeable. *Right panel:* The second type of cell-to-cell communication, also referred to as direct, includes interactions mediated by cell adhesion molecules and the cytoskeleton, as well as the diffusional exchange of cytoplasmic ions and molecules through gap junction channels

excellent reviews [7–9]. I shall then discuss in more detail the evidence pointing to the participation of direct β-to-β-cell communication in the control of insulin secretion.

Intercellular Communication in the Endocrine Pancreas

Communication Between Islets of Langerhans

Insulin secretion is a multicellular event, resulting from the coordinated activation of numerous β-cells, which form the bulk of distinct islets of Langerhans. Hence, some mechanism should functionally link these functional subunits that collectively form the endocrine pancreas. Furthermore, since insulin secretion normally undergoes cyclic fluctuations [19,20] and should be continuously adapted to the changing needs of the organism, the mechanism coordinating the functioning of individual islets should also allow for a dynamic, moment-to-moment modulation of β-cell functioning.

Coordination between β-cells forming distinct islets must rely on mechanisms of indirect cell-to-cell communication, since the islets of Langerhans are separated from each other by basal laminae, connective tissue constituents and exocrine acini. A large variety of hormones and neuromediators modulate insulin secretion under conditions thought to have physiological relevance [7,8,9,19]. Experimentally, stimulation of insulin secretion is induced by gastric inhibitory peptide, thyrotropin-releasing hormone, glucagon-like peptide I, β-adrenergic agonists, acetylcholine, cholecystokinin, gastrin-releasing peptide and vasoactive intestinal peptide. Conversely, inhibition is induced by corticotropin-releasing factor, peptide YY, atrial natriuretic peptide, pancreastatin, α_2-adrenergic agonists, galanin and neuropeptide Y. However, the factors which actually control insulin secretion in vivo remain to be clearly elucidated, as does their relative contribution to β-cell control. The current view is that the secretory response of β-cells, which is initiated by circulating nutrients, is essentially modulated by gut hormones, and particularly by gastric inhibitory peptide and cholecystokinin [7,8,9,19].

Clearly, however, no physiological stimulus increases the levels of glucose, other nutrients and gut hormones with the speed and magnitude required to induce an almost immediate and synchronous activation of all pancreatic islets. Such a massive, all/none signal may in fact be inappropriate to ensure the long-term pulsatility of insulin secretion, which is observed in vivo and in vitro [19,20]. Hence, fine tuning of islet coordination requires an additional connection between these microorgans. The observation that pulsatility of insulin secretion is preserved in the absence of hormonal circulation [19,20] suggests that this connection is ensured by the intrapancreatic nervous system. Since insulin pulses are perturbed by tetrodotoxin but not by drugs blocking cholinergic and adrenergic receptors, it is thought that islet coordination involves autonomous intrapancreatic ganglia which are likely to serve both as pacemakers and as integration centres [19]. These ganglia receive adrenergic, cholinergic and peptidergic inputs, and their postganglionic fibres are seen to interconnect endocrine islets [9].

Communication Within Islets of Langerhans

Under in vitro conditions, isolated islets of Langerhans retain most of the secretory characteristics of the intact pancreas and, in particular, still respond acutely to the natural stimuli of the gland, including glucose. Albeit quantitatively reduced, the pattern of insulin release and biosynthesis observed during this stimulation is analogous to that observed in situ, indicating that islets actually represent competent functional subunits of the endocrine pancreas. Because each islet releases amounts of insulin that largely exceed those made up by individual β-cells, proper functioning of an islet implies an adequate coordination between the β-cells which form its core. This coordination is likely to be achieved by multiple mechanisms, involving both indirect and direct intercellular communication.

Indirect β-Cell Communication

Individual islets of Langerhans respond to most of the neuromediators and hormones which affect the endocrine secretion of the intact pancreas, including the four main hormones produced by the islets themselves [7,8,21]. Thus, insulin inhibits glucagon secretion, and possibly also the secretion of somatostatin, pancreatic polypeptide and insulin itself; glucagon stimulates insulin and somatostatin secretion; pancreatic polypeptide inhibits somatostatin secretion, and somatostatin inhibits the release of all islet hormones [7,8,21]. Some of these effects can be prevented by antibodies to specific hormones, suggesting that they may well be elicited in vivo after release of endogenous islet products [7,8,21]. Since islet cells are influenced by hormone levels much lower than those found in the venous effluent of the pancreas [7,8,21], it is unlikely that any putative paracrine effect merely occurs by diffusion of hormones from the producing cell to nearby targets through a continuous extracellular space. If this were to be the case, it is likely that islet cell receptors would be downregulated by the sustained exposure to high hormone concentrations. However, it is certainly possible that islet hormones be channelled to specific membrane domains, either by restricted diffusion in the extracellular islet space or by vectorial transport in the islet microcirculation [7,8,19,21]. Microvessels originate from afferent arterioles in the centre of the islets, which predominantly comprise insulin-producing β-cells, and direct blood flow to the islet periphery [22], where β-, α-, δ- and PP-cells are intermixed [23]. Comparison of arterial and venous infusions in the presence and absence of hormone-neutralizing antibodies suggests that blood flow reaches β-cells first, then α-cells and finally δ-cells [21,22]. According to this scheme, high concentrations of insulin should almost continuously bathe the peripheral islet mantle, whereas glucagon and somatostatin would have little chance of reaching most β-cells without first entering systemic circulation. However, whether and how islet hormones actually have any paracrine effect in vivo is still the subject of substantial debate, since we still do not know the actual concentration of hormones in the interstitial islet fluid, the flux and direction of this fluid and the distribution of islet cell receptors.

Autonomic ganglia have not been observed within isolated islets [9], indicating that basal and stimulated insulin secretion may be sustained in the absence of innervation. Nevertheless, interruption of neural input modifies islet function, as revealed by an increase in the frequency of the oscillations of insulin secretion [20]. The rapid oscillations observed within individual, isolated islets are probably driven by periodic fluctuations in the levels of intracellular free Ca^{2+} and/or in the glycolytic activity of β-cells [24]. In vivo, these oscillations may be masked by the longer cycle pulses that result from the neural coordination of multiple islets.

In addition to hormones and neuromediators, numerous other signals flowing through extracellular spaces may coordinate β-cell behaviour. For example, it has been noted that electrophysiological characteristics are quite variable at the level of single β-cells and become more uniform and stable when β-cells form clusters [24–26]. Within intact islets, virtually all β-cells show a high degree of electrical synchronization during both silent and burst periods of electrical activity [24–29], implying an almost immediate intercellular coordination of the levels of current-carrying ions. This coordination has been verified for Ca^{2+}, whose oscillatory levels appear to be determined collectively by groups of synchronized β-cells rather than by individual β-cells [24], and may rapidly equilibrate across whole islets [30,31]. The mechanism responsible for this electrical synchronization remains to be elucidated. Even though it is likely that gap junctional coupling is involved (see below), other mechanisms may also contribute. For example, the occurrence in the extracellular islet fluid of K^+ waves, which regularly precede the bursts of electrical activity [32], may provide a rapid and efficient way for coordinating membrane depolarization, as well as the subsequent Ca^{2+} oscillations and insulin release in distant β-cells [10].

Direct β-Cell Communication

The secretory function of β-cells is preserved in vitro under conditions which perturb the innervation, blood supply and flux of extracellular fluid that normally interconnect pancreatic cells in vivo. However, maintenance of regulated secretion is dependent on preservation of at least some of the contacts β-cells establish within the islets of Langerhans. Disruption of such contacts markedly perturbs the secretory response of β-cells [1,11,13–18]. This observation and the rapid recovery of a qualitatively normal secretory response after reestablishment of intercellular contacts suggest an important regulatory role for direct cell-to-cell communication. Of the several forms of interaction which may operate between cells in contact, only the intercellular exchange of ions and molecules via membrane channels made of connexins has so far been studied in any detail for its possible implication in the control of insulin secretion [10].

Connexin proteins form channels which cluster on specialized regions of the cell membrane called gap junctions. These membrane domains are present on virtually every cell type of multicellular organisms throughout the phylogenetic scale [3,6,33,34]. In particular, gap junctions and coupling appear to be obligatory attributes of endocrine and exocrine secretory cells, including pancreatic β-cells [35]. Gap junction channels are characterized by a high conductance and permeability, span the membrane of adjacent cells and extend across the extracellular space [3].

Within this space, the channels of one cell dock with the channels of an adjacent cell, so that the extracellular space is bridged by a continuous, two-cell hydrophilic pathway. It is through this pathway that ions and molecules of less than 900 dalton directly pass from the cytoplasm of one cell into the cytoplasm of an adjacent cell, without entering the extracellular compartment. This passage is also referred to as ionic and metabolic (cell-to-cell or junctional) coupling [3].

β-Cell Communication with Connexins

Characteristics

Pancreatic β-cells are connected to each other by numerous and minute gap junctions [1]. These structures are made of a 43-kDa protein, referred to as Cx43, which is part of the connexin family [35,36]. In a resting β-cell, there are 800–2000 gap junction channels that, altogether, occupy less than $1\,\mu m^2$ of the plasma membrane [37–39]. These channels allow for the β-to-β exchange of current-carrying ions, as demonstrated by the intercellular spread of electrotonic potential changes induced by pulsing currents into one cell [27,28], by the detection of junctional conductances at individual β-to-β-cell interfaces [36,40], and by the synchronization of free Ca^{2+} oscillations [24,30,31] and membrane potentials across the islets [27,29]. Gap junction channels also mediate the metabolic coupling of β-cells, as shown by the rapid β-to-β-cell exchange of several molecules which do not permeate the cell membrane, such as exogenous fluorescent probes, nucleotides and glycolytic intermediates [36,41–45].

Assuming that the conductance of Cx43-made channels is 50 pS [3], β-cells could be electrically coupled by only one to two fully opened channels. If, as in other systems, these channels are opened for about 10% of the time, β-cells would need 10–20 channels to directly exchange current carrying ions. This figure is consistent with the size of β-cell gap junctions, as revealed by freeze-fracture electron microscopy [37,39]. Ionic and metabolic coupling are not observed between all β-cells [36,40–44], in agreement with the detection of gap junctions in about 30% of the β-to-β-cell interfaces [39]. Furthermore, metabolic coupling is restricted to small groups of β-cells within both intact islets of Langerhans and monolayer cultures [36,41–44]. Hence, the 10–6000 β-cells that are found in a pancreatic islet appear to be functionally grouped in multiple communication territories [10,46] rather than in a single syncytial unit [23].

Several lines of evidence indicate a contribution of β-cell coupling to the control of insulin secretion. Firstly, single β-cells are either totally unresponsive to glucose and other stimuli, or show a perturbed secretion, as indicated by increased basal release of insulin, poor responsiveness to secretagogues, decreased protein biosynthesis, decreased basal expression of the insulin gene and loss of its normal, cyclic adenosine monophosphate (cAMP) dependent control [11–18,47,48]. Restoration of β-to-β-cell contacts is paralleled by a rapid improvement in secretion, as judged by decreased basal release of insulin, increased responsiveness to secretagogues, protein biosynthesis and expression of the insulin gene [11–18,47,48]. Even though it is clear that gap junctions may rapidly form during β-cell reaggregation [12,13], there is as yet no definitive evidence that junctional coupling is the sole event accounting for these secretory changes. Nevertheless, the finding that a drug which blocks gap junction channels prevents the increase in insulin secretion that normally occurs as single β-cells interact to form pairs [18,49] indicates that coupling may acutely control the recruitment of secreting β-cells as well as their individual level of activity.

Secondly, in vivo as well as in vitro, sustained stimulation of insulin release increases β-cell coupling [36–39,46] due, at least in some cases, to increased expression of gap junctions and Cx43 [36]. As yet, there is much less evidence for a more acute regulation of β-cell coupling resulting from changes in the conductance and/or permeability of existing gap junction channels. While this possibility is suggested by the almost immediate increase in coupling coefficient and in the synchronization of both electrical activity and Ca^{2+} oscillations, observed when islets are exposed to concentrations of glucose stimulating insulin release [28,30], it remains to be determined to what extent these changes are actually due to improved junctional communication.

Thirdly, conditions that inhibit insulin release decrease or abolish β-to-β-cell coupling [44,46]. In vivo, however, an opposite change may be seen whenever the inhibition of β-cell secretion results in hyperglycaemia [46]. It is therefore possible that the level of circulating glucose and the ability of β-cells to properly recognize the sugar stimulus may independently influence junctional coupling.

Fourthly, the acute pharmacological blockade of gap junction channels markedly alters β-cell function, as indicated by increased basal insulin release and loss of stimulated insulin secretion from both isolated islets of Langerhans and intact pancreas [49,50]. These alterations, which could not be accounted for by concomitant changes in the main second messengers controlling insulin secretion, were rapidly and fully reversible after wash-out of the uncoupling drugs and were not observed in single β-cells [49], indicating that they may well be a specific result of cell uncoupling.

The fifth line of evidence indicates that a number of tumoral and transformed insulin-producing cell lines which show abnormal glucose sensitivity do not express connexins and gap junctions and are essentially uncoupled [51]. These defects cannot be explained simply by the in vitro down-regulation of gap junctional channels since primary β-cells in monolayer cultures continue to express Cx43, gap junctions and coupling [42,51]. Nor can these defects be ascribed solely to the proliferation capacity of the cell lines, since the glucose-sensitive cells of a rat insulinoma, from which at least two of the communication-deficient lines were derived, still express Cx43, gap junctions and coupling in vivo [41,51]. Hence, coupling defects appear to be a shared attribute of cell lines showing abnormal sensitivity to glucose, whatever their expression of several other key factors (GLUT-2 glucose transporter, glucokinase, hexokinase, etc) that rate-limit the entry, the metabolism and the early secretory response to the sugar.

Finally, a coupling–secretion relationship is supported by experiments on cell lines that had been stably transfected with the gene coding for Cx43 and expressed gap junctions and coupling similar to primary β-cells. As compared to wild-type, noncommunicating controls, the transfected cells displayed enhanced expression of the insulin gene, a markedly increased insulin content, and preserved the ability to increase insulin release when stimulated by physiologically relevant concentrations of glucose [51]. In contrast, several clones of the same cells which expressed unusually high levels of Cx43, gap junctions and coupling did not show any improvement of these secretory characteristics. Actually, most of the latter cells showed loss of the major features which characterize primary β-cells [51].

These observations indicate that an adequate expression of Cx43 and of gap junctional communication is required for proper control of both insulin biosynthesis and secretion [10,48,51].

Why Communication Is Needed

The studies summarized above demonstrate a close parallel between junctional communication and the ability of β-cells to appropriately release and biosynthesize insulin. They also raise the question of why such communication is required for proper β-cell functioning. Some answers to this question have been derived from studies comparing single β-cells, which cannot form functional gap junctions, to (re)aggregated β-cells that express these structures.

All/None Response of Single β-Cells

Haemolytic plaque assays have revealed that individual β-cells secrete insulin in a rather variable way after stimulation by glucose, nonglucidic nutrient secretagogues and other nonmetabolizable stimuli [17,18,47,52,53]. Thus, irrespective of the type and duration of the stimulation, secreting β-cells were seen to consistently coexist with apparently similar β-cells, which did not release detectable amounts of the hormone, if they secreted at all. These nonsecreting cells were shown to exclude trypan blue, to display an electrophysiologically normal membrane and, at least in some cases, to be capable of new protein biosynthesis [17,18,47,52,53], indicating that they were living and functioning properly. Most secreting and nonsecreting β-cells were also observed to maintain their different secretory behaviour during successive stimulations and for various intervals of secretagogue challenge [54], suggesting that they probably represent functionally distinct and quantitatively stable subpopulations of β-cells. Successive stimulations of the same β-cells by different secretagogues further indicated that β-cells unresponsive to an acute glucose stimulation, were also usually unresponsive to other secretagogues [52,53].

By evaluating the proportion of cells forming a haemolytic plaque, a parameter which estimates the fraction of β-cells contributing to insulin output, and the individual areas of these plaques, which evaluate the amount of hormone released by each secreting cell, only a small fraction of β-cells was found able to release detectable levels of insulin in the presence of 2.8 mM glucose [17,55]. Raising the concentration of D-glucose in the medium increased the proportion of secreting β-cells up to a maximum of about 40% (a tenfold increase as compared to basal value) in the presence of 16.7 mM glucose. Under these maximally stimulatory conditions, the areas of individual haemolytic plaques were again variable from one cell to another, and, on average, only slightly elevated over those observed under basal conditions [17,55]. Comparable observations were made with β-cells cultured for 15–18h in the presence of 2.8 mM glucose and then exposed for 30 min to a stimulatory concentration (16.7 mM) of the sugar. Under these conditions, overall insulin release was markedly decreased due to a significant reduction in the proportion of cells that were recruited to secrete, but the amount of insulin which was released by individual cells was unaltered [17,55]. These observations indicate: (1) that secreting β-cells do not all release similar amounts of insulin, whatever the glucose concentration prevalent in the medium; (2) that the amount of insulin released by glucose-sensitive cells is not markedly influenced by the level of stimulation; (3) that the increased insulin secretion which is elicited by glucose is mostly due to an increased recruitment of secreting cells; (4) that under conditions which seriously hinder glucose-induced insulin release, some β-cells still secrete the hormone in amounts comparable to those observed under conditions which optimize secretion. Hence, β-cells appear to show an all/none type of secretory responsiveness. When the total insulin output of these cells is computed as a function of glucose concentration, a sigmoidal curve is obtained which is similar in shape to that observed for the glucose-induced insulin secretion of intact pancreas [55,56]. Thus, in spite of a marked individual heterogeneity, single β-cells show, as a population, a response pattern which is qualitatively normal.

β-Cell Heterogeneity

The reason why live and apparently similar β-cells feature the marked secretory heterogeneity summarized above is still a matter of conjecture. Comparison of secreting and nonsecreting β-cells, as differentiated by plaque assay, failed to detect any difference in their expression of both GLUT2 and glucokinase. As judged by immunofluorescence staining with specific antibodies, these two proteins were expressed by both glucose-sensitive and-insensitive cells [52]. In addition, electrophysiological measurements showed that secreting and nonsecreting cells had similar ion channels for voltage-dependent Na^+ currents, voltage-activated Ba^{2+} currents, voltage-dependent K^+ delayed-rectifier currents, voltage-dependent Ca^{2+}-activated K^+ currents, and voltage-independent and tolbutamide-sensitive K^+ currents and showed a similar modulation of some of these conductances during both glucose and tolbutamide stimulation [53]. Eventually, secreting and nonsecreting cells were found capable of comparable changes in the reduced form of nicotinamide adenine dinucleotide phosphate (NAD(P)H) autofluorescence after an acute glucose stimulation, even though the subpopulation of β-cells featuring a glucose-induced increase in the NAD(P)H signal secreted more insulin than the subpopulation showing no apparent redox shift [57]. However, the average size of β-cells differed in the two sorted populations [57,58]. When cells were size-matched, no further difference in insulin output was found between the β-cells that appeared to be metabolically active and those which seemed metabolically quiescent [57]. Hence, the secretory heterogeneity of β-cells appears related to a difference in size rather than to a marked differential ability to activate metabolism in the presence of glucose.

Parallel experiments have shown that the same β-cell subpopulation responds to D-glucose and to nonglucidic nutrient secretagogues, and do so to the same extent, as judged both by a similar total insulin output and by a comparable number of cells contributing to it [52]. These findings strongly suggest that the factor(s) responsible for the responsiveness of individual β-cells are not tightly related to a specific metabolic event in the early steps of glucose transport and catabolism. Functional heterogeneity was also observed in the presence of nonmetabolized secretagogues [17,53], further indicating that the factors which rule the response of β-cells are likely to be somewhat distal in the sequence of steps which link stimulus to secretion.

Eventually, autoradiographic studies have revealed that individual β-cells are also heterogeneous in terms of their ability to biosynthesize proteins, including insulin, and that glucose stimulates protein biosynthesis via a dose-dependent recruitment of β-cells which is similar to that observed for secretion [59]. These observations argue for the same factors controlling the secretory and biosynthetic ability of β-cells. By simultaneous monitoring of insulin secretion and protein biosynthesis, β-cells which were biosynthetically active were found to be preferentially recruited for insulin secretion during a short-term stimulation by glucose, as compared to similar, but biosynthetically quiescent β-cells [47]. The additional observation that a sizable proportion of the former cells was unable to secrete insulin and, conversely, that a number of biosynthetically quiescent β-cells could release detectable amounts of the hormone means it is unlikely that a major metabolic perturbation accounts for the unresponsiveness of some β-cells to glucose.

How Communication May Work

The above findings indicate that individual β-cells show a marked functional heterogeneity, the cause of which remains uncertain. They also indicate that, in spite of this heterogeneity, single β-cells, as a population, have a response pattern which is qualitatively similar to that observed in the whole pancreas. However, as compared to the situation prevailing in the intact gland, the quantitative response of single cells to stimulation is severely reduced. This reduction is essentially due to a rather modest recruitment of single cells able to release and to biosynthesize insulin during secretagogue stimulation. Hence, some intra-islet mechanism should enhance the recruitment of functioning β-cells under conditions requiring substantial insulin output and biosynthesis. Because such a mechanism is apparently lost in single cells, we investigated whether it could be related to the establishment of cell-to-cell contacts.

Contact-Dependent Improvement of Insulin Secretion

Comparison of intact islets, isolated islet cells and clumps of (re)aggregated cells has consistently shown that single β-cells (which cannot form gap junctions) are poorly responsive to glucose and rapidly recover an almost normal responsiveness to the sugar after reaggregation and reestablishment of junctional contacts [11–18]. The main difference is that stimulatory concentrations of glucose elicit secretion from only a small proportion of single β-cells, whereas they progressively recruit for secretion much higher numbers of β-cells in contact [14,17,18,47,53,55].

As a first approach to investigate this effect, single β-cells were compared to β-cell pairs, the smallest units in which cell-to-cell contact may be influential. Insulin release of both single and paired β-cells was found to increase as a function of the concentration of glucose prevalent in the medium [18,55]. However, whereas in the presence of glucose concentrations below 5.7 mM, single β-cells and β-cell pairs behaved similarly, this was no longer the case in the presence of higher concentrations of the sugar. Under conditions of medium to maximum stimulation, β-cell pairs secreted about twice as often as single β-cells and also released on average twice as much insulin [18].

From these changes, it can be calculated that pairs of β-cells secrete four times more insulin than single β-cells and, hence, that on average, individual β-cells function twice as much when they are paired to another β-cell than when they are single. The data are consistent with the possibility that only one β-cell contributes to secretion in pairs exposed to low glucose levels, whereas the two contacting β-cells do so in coordination within pairs exposed to high glucose levels. These results indicate, first, that under basal conditions the mere contact between two β-cells is not sufficient to promote insulin release. They also suggest that a number of β-cells which are quiescent may be induced to secrete after contact with another β-cell. Eventually, they indicate that the mechanism which mediates this functional recruitment is glucose-dependent [55].

To gain further insight into this mechanism, we investigated the time required to observe the fourfold increase in glucose-induced insulin release which takes place as single β-cells adhere to form pairs. To this end, single β-cells were prelabelled with inert fluorescent beads, which are conveniently stable cytoplasmic markers, and were cocultured with similar but unlabelled β-cells. Using this approach, we established that within 10 min of coculture, β-cell pairs already showed the full increase in insulin secretion which is observed in much older pairs [55]. Thus, even though cell contact is not obligatory for insulin secretion, at least for some β-cells it acutely controls the number of secreting β-cells and their individual level of activity.

To assess whether junctional coupling is the contact-dependent mechanism involved in this control, insulin secretion was studied under conditions of pharmacological gap junction blockade. Using drugs which block gap junction channels [49,50,60], possibly by interfering with the normal phosphorylation of Cx43 [61], a reversible and marked alteration of insulin secretion was found to parallel changes in β-cell coupling within intact islets [49]. Thus, under conditions abolishing this coupling, the basal release of insulin increased and the islets lost their responsiveness to glucose and other secretagogues in the absence of detectable alterations in cytosolic free Ca^{2+}, pH and in the Ca^{2+}-stimulated insulin release of electropermeabilized cells [49]. These data indicate that the uncoupling drugs did not pleiotropically perturb the secretory machinery of β-cells, as also indicated by the full reversibility of the secretion changes within minutes of the drug washout and restablishment of normal β-cell communications [49]. Furthermore, the uncoupling agent heptanol had no effect on the secretion of single β-cells, while it markedly reduced that of β-cell pairs [49]. Under these conditions, paired β-cells secreted, on a per cell basis, as single cells, even though they preserved a close appositional contact [49]. It remains to be shown whether these secretory changes were caused solely by interruption of β-to-β-cell communications. Future studies using antisense mRNA strategies and/or anticonnexin antibodies should help to provide a conclusive answer to this question. At any rate, the available findings point again to a close parallel between the acute control of insulin secretion and that of β-cell coupling.

Possible Physiological Relevance

To date there is little evidence to suggest that what we have learned in vitro may also be relevant to the physiological conditions prevailing in vivo. However, gap junctions have been identified within in situ fixed islets [1], and have been shown to be more numerous in the centre than at the periphery of these microorgans [38]. Interestingly, the β-cells located in the centre of the islets were found to be larger and to release and biosynthesize insulin at a higher rate in response to glucose than the β-cells located at their periphery [55,62]. Hence, in vivo as well as in vitro β-cells show a striking heterogeneity in size and in their ability to release insulin and biosynthesize proteins. Furthermore, both in vivo and in vitro, this heterogeneity is most easily revealed after glucose stimulation. It is therefore likely that the β-cells located in the centre of the islets may contribute predominantly to insulin homeostasis under most physiological conditions. The cause of this regional, intra-islet heterogeneity remains to be demonstrated. Differences in junctional coupling [45], as well as differential interactions of central and peripheral β-cells with hormones, neurotransmitters and other signals carried by circulation could certainly contribute to it.

A Model

The observations summarized above provide compelling evidence that a mechanism dependent on the physical contact of β-cells ensures proper glucose responsiveness and upregulates secretagogue-induced insulin release, protein biosynthesis and the expression of the insulin gene. Conceivably, this mechanism may involve any of the events that take place as β-cells interact with one another, including changes in the polarity of cytoplasmic and membrane components, and in the surface molecules which ensure cell adhesion and form intercellular junctions. However, the observation that, under stimulatory conditions, a sizeable fraction of β-cell aggregates does not release insulin and synthesize proteins indicates that mere aggregation is not sufficient to ensure an upregulation of β-cell functions. Furthermore, since the phenotype of nonsecreting β-cells in similar to that of secreting β-cells, it is unlikely that changes in the polarity of cytoplasmic components and/or in the molecules and structures which ensure β-cell adhesion are prime events in the contact-dependent regulation of insulin secretion and biosynthesis. One implication of this view is that another mechanism, activated by cell contact, should be prominent.

Several lines of experimental evidence, including the rapid up-regulation of secretion which takes place after β-cell aggregation, suggest that such a control mechanism could be provided by junctional coupling. However, the reason why coupling may be required, if not obligatory, for proper β-cell functioning remains to be elucidated, as does the molecular mechanism underlying this requirement.

The existence of a substantial heterogeneity of individual β-cells, which result in widely different abilities to release and biosynthesisze insulin, provides a first conceptual lead for understanding why coupling may be advantageous to β-cells [10]. Gap junctional communication is a most straightforward way to balance and coordinate differences between cells [3]. In such a system, the increase of cytoplasmic ions or molecules less than 900 dalton into one cell is followed by their diffusion-driven passage into nearby cells (Fig. 2). At steady state, this passage leads to the equilibration of electrochemical concentrations on both sides of the channels. If the resulting concentration reaches a threshold level for activation of an effector mechanism, functioning will be modified not only in the cell in which the ionic and molecular change first occurred, but also in all other cells coupled to it. In this way, junctional coupling may ensure the functional recruitment of cells that could not be directly activated. Hence, it is possible that disparities in the concentrations of ions and metabolites which are essential for secretion may cause individual β-cells to asynchronously biosynthesize and release insulin (Fig. 2). By equilibrating ionic and molecular electrochemical gradients between intercommunicating cells, junctional coupling could correct these disparities. As a result, β-cells deficient in some critical factor could gain it from a neighbour β-cell [63] and, once a concentration threshold is reached, be synchronously activated with nearby cells (Fig. 2). This implication is consistent with the observations that coupling promotes the recruitment of secreting and biosynthetically active β-cells [18,49], and synchronizes Ca^{2+} changes between β-cells in contact [24,30,31,64,65]. Such a synchronization would certainly be advantageous whenever large amounts of insulin are needed, e.g., after meals or exercise. Conversely, blockade of this synchronization due to closure of junctional channels

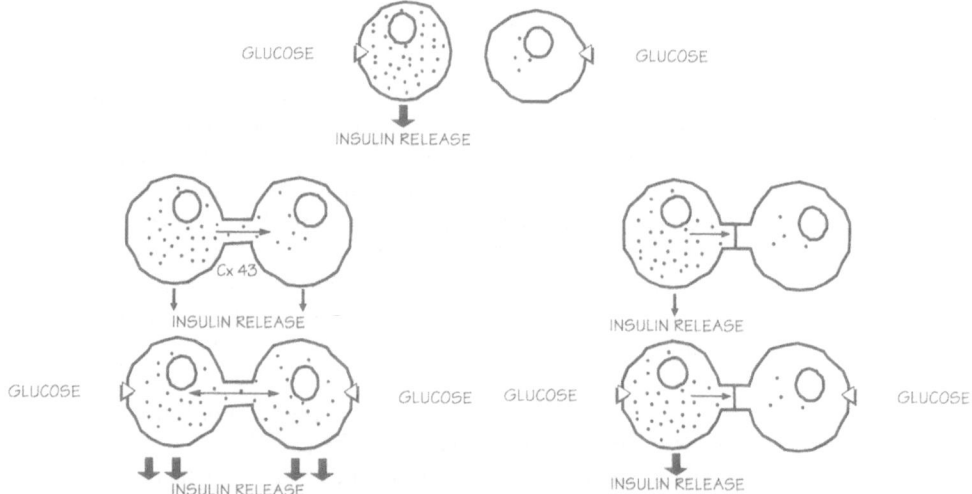

Fig. 2. Gap junctional communication and insulin secretion. The experimental observations so far made on primary insulin-producing cells and lines derived thereof, may be summarized by the following model. *Upper panel.* Natural secretagogues, including glucose, directly stimulate only a proportion of β-cells in which the concentrations of critical ions and molecules (represented by the *black dots*) reach appropriate threshold levels. *Lower left panel.* Establishment of gap junctional communication permits the diffusion-driven passage of these ions and molecules from secreting into non-secreting cells. As a result, the latter cells become activated and synchronized with those already functioning, even in the absence of external stimuli. Addition of secretagogues further increases the response of gap junction-sharing cells, presumably because the rapid equilibration of ionic and molecular concentrations across junctional channels optimizes the threshold level for activation. *Lower right panel.* Conditions which result in the acute and temporary blockade of gap junction channels prevent this equilibration. As a result, uncoupled β-cells display a secretory responsiveness which is as heterogeneous as that of single cells. They also secrete less insulin than coupled β-cells, in spite of structurally similar intercellular relationships

could ensure the rapid switch-off of most β-cells, which prevails under basal conditions. Furthermore, since on an individual basis β-cells secrete more insulin when they share gap junctions than when they are single [18,55], the coupling-induced equilibration should also optimize the concentration of factors critical for secretion. This view is consistent with the observation that the threshold level for activation of insulin secretion and biosynthesis is lowered under conditions promoting β-cell coupling [13,14,67].

The molecular mechanism whereby these effects may be achieved is still speculative. A first possibility is that coupling enhances the secretagogue-induced changes in free intracellular Ca^{2+}. Indeed, the levels and oscillations of this cation, which plays a critical role in the control of insulin secretion, are differentially affected by the establishment of contacts between β-cells [24,64]. Second, coupling may modulate the electrophysiological characteristics of the β-cell membrane. It has been observed that resting membrane potentials, as well as individual conductances, become larger and more stable after establishment of cell-to-cell contacts, presumably because individual β-cells are equipped with such limited numbers of K^+ and Ca^{2+} channels that they cannot ensure stable ionic fluxes, as these channels fluctuate between the open and closed states [68]. By mediating the intercellular equilibration of current-carrying ions, junctional coupling may result in the functional sharing of these channels, which would then be available in sufficient numbers to all coupled cells. Channel sharing could markedly promote the responsiveness of β-cells, since the activation of the distal, effector steps of the secretory machinery depends on proper control of membrane potentials and ionic fluxes. Clearly, intercellular exchanges of both Ca^{2+} and K^+ could also account for the electrical synchronization of β-cells which has been observed in intact islets of Langerhans [27–29]. Third, coupling could ensure the spreading of signals controlling secretion across large cell populations. Secretagogue-induced Ca^{2+} waves, indicating a temporally and spatially coordinated change in the levels of this cation, have been observed throughout intact islets [30,31,65]. These waves may result from the intercellular exchange of Ca^{2+}, or IP_3, via gap junctions and may mediate the rapid recruitment of secreting cells at a distance from the site of signalling. An analogous mechanism has been shown to ensure the hormonal stimulation of cells devoid of cognate receptors, provided these cells were coupled to adjacent cells able to recognize the signal [69–71]. This effect, which is probably due to a gap junction-mediated transfer of second messengers, may operate in the islet to functionally recruit β-cells deficient in factors which rate-limit secretion. By sharing gap junctions with cells rich in these essential factors, the defective cells may be adequately activated to secrete, even though, individually, they may be unable to directly recognize stimuli. At least in vitro, such an activation can be demonstrated for glucose-induced insulin release [55].

Direct β-Cell Communication and Diabetes

The complex interplay of multiple forms of direct and indirect communication which control β-cell function and the major influences these communications have on insulin release and biosynthesis raise the possibility that in vivo perturbations in the communication network interconnecting β-cells are relevant to the early pathogenesis of diabetic disorders [72,73]. To date, most attention has been paid to defects in mechanisms of indirect cell-to-cell communication. Hence, type II diabetes, for example, has been shown to be associated with, if not preceeded by, alterations in nutrient, neural and hormonal signalling to β-cells [7,19,73]. However, our understanding of whether any of these changes significantly contribute to abnormal β-cell function in the early steps of the disease remains at best limited.

An implication of the close relationship which exists between junctional communication and secretion is that abnormal secretion patterns may also result when direct cell-to-cell communication is perturbed. As yet, this possibility has not been directly investigated, even though it is supported by a number of circumstantial observations. First, the absence as well as the blockage of β-cell gap junctions is associated with increased basal insulin release and loss of responsiveness to glucose [11–18,49], two secretory defects which characterize type II diabetes [72–74]. Second, several lines of insulin-producing cells which feature selective defects in glucose responsiveness and decreased insulin content do not express detectable connexins, gap junctions or coupling [51]. Third, glibenclamide, a sulphonylurea which improves insulin secretion in type II diabetics, improves the expression of Cx43, gap junctions and coupling between β-cells [36–39,46]. It is therefore possible that β-cell connexins and coupling are defective in type II diabetes and may be implicated in the initial and still obscure, derangements of intercellular signalling which initiate this chronic disease. Biochemical testing of this implication can now begin in humans, using antibodies and probes to evaluate the expression of connexins and their specific transcripts.

Conclusions and Perspectives

Insulin secretion by the endocrine pancreas is the result of the activity of numerous and functionally heterogeneous cells whose integration necessarily depends on proper communication. As the mechanisms of this communication are unravelled, it becomes evident that full control of β-cell function cannot be ascribed solely to one regulatory mechanism, but rather to the interaction between multiple mechanisms that involve cell-to-cell signalling by nutrients, intrinsic and extrinsic neural inputs, local and circulating hormones, as well as direct interactions between adjacent cells. A major problem in unravelling the regulation of insulin secretion stems from the complexity of the interactions between these signals which converge at the islet level. Most likely, the multiplicity of mechanisms for intercellular coordination permits β-cells to properly adapt their hormonal secretion to the needs of the organism, which continuously change throughout the day, depending on the physiological condition. It also provides these cells with a multistage feedback system, which, if somewhat redundant and costly, may be instrumental in preserving an adequate secretory function under pathophysiological conditions, resulting in perturbation of one or more systems for cell communication. Several of the mechanisms ensuring intercellular communication overlap and, hence, may partly or totally replace each other in case one mechanism is disturbed.

The precise contribution of different communication mechanisms to the net response of pancreatic islets remains to be fully understood, as does their hierarchic organization. In this respect, the finding of a gap junction-mediated modulation of β-cell functions, under conditions abolishing indirect cell-to-cell communication, indicates a fundamental, hitherto disregarded role of cell-to-cell coupling. In view of the functional heterogeneity of β-cells, it is probable that this mechanism has become an obligatory feature since it provides the most direct way for compensation of intrinsic metabolic and effector differences of secretory cells. By equilibrating ions and molecules across domains of coupled cells, junctional communication has the potential to drastically alter multiple aspects of the secretory function, and may be essential for building an appropriate output of insulin, starting from a highly heterogeneous β-cell population.

At this stage, we know that junctional communication is tightly linked to the ability of β-cells to appropriately release and biosynthesize insulin and that defective coupling may participate in the loss of β-cell responsiveness to secretagogues. However, the molecular mechanism underlying the relationship between β-cell coupling and secretion remains to be unravelled, and our understanding of whether such a relationship is causal is limited by several unknowns. Hence, the reasons why gap junctional coupling should be implicated in secretion, and the way this implication is achieved, remain essentially speculative. It also, remains to be assessed whether and how coupling defects participate in the early and still obscure pathogenesis of pancreatic dysfunctions of major relevance to human medicine. As yet, such a possibility has not been investigated. The recent availability of novel cell and molecular biology tools, as well as of novel strategies with which to interfere with specific steps of junctional coupling and secretion, now offers the exciting perspective of directly addressing these questions, in vitro as well as in vivo.

Acknowledgments. Most of the studies from my laboratory were performed by Drs. D. Bosco, M. Chanson, A. Oeggerli, D. Salomon, C. Vozzi, and T. White, to whom I am very grateful. I also thank N. Bolzonello, L. Burkhardt, A. Charollais, P. Fruleux, J.-P. Gerber, F. Gros and E. Sutter for excellent technical assistance. The work of this team was supported by grants from the Swiss National Science Foundation (32-043086.95), the International Juvenile Diabetes Foundation (195077) and the Commission of the European Union (BMH4-CT96-1427).

References

1. Orci L, Unger RH, Renold AE (1973) Structural coupling between pancreatic islet cells. Experientia 29: 1015–1018
2. LeRoith D (1990) Are all cells "endocrine"? In: Becker KL (ed) Principles and practice of endocrinology and metabolism. Lippincott, Philadelphia, pp 10–13
3. Bennett MVL, Barrio LC, Bargiello TA, Spray DC, Hertzberg E, Saez JC (1991) Gap junctions: new tools, new answers, new questions. Neuron 6: 305–320
4. Edelman GM, Crossin KL (1991) Cell adhesion molecules: implications for a molecular histology. Annu Rev Biochem 60: 155–190
5. Greenwald I, Rubin GM (1992) Making a difference: the role of cell-to-cell interactions in establishing separate identities for equivalent cells. Cell 68: 271–281
6. Beyer EC, Paul DL, Goodenough DA (1990) Connexin family of gap junction proteins. J Membr Biol 116: 187–194
7. Samols E, Stagner JI (1991) Intraislet and islet-acinar portal systems and their significance. In: Samols E (ed) The endocrine pancreas. Raven, New York, pp 93–124
8. Marks V, Samols E, Stagner J (1992) Intra-islet interactions. In: Flatt PR (ed) Nutrient regulation of insulin secretion. Portland, London, pp 41–57

9. Berggren PO, Rorsman P, Efendic S, Ostenson CG, Flatt PR, Nilsson T, Arkhammar P, Juntti-Berggren L (1992) Mechanisms of action of entero-insular hormones, islet peptides and neural input on the insulin secretory process. In: Flatt PR (ed) Nutrient regulation of insulin secretion. Portland, London, pp 289–318
10. Meda P (1995) Junctional coupling of pancreatic β-cells. In: Huizinga JD (ed) Pacemaker activity and intercellular communication. CRC, Boca Raton, pp 275–291
11. Pipeleers D (1984) Islet cell interactions with pancreatic B-cells. Experientia 40: 1114–1126
12. Chertow BS, Baranetsky NG, Sivitz WI, Meda P, Webb MD, Shih JC (1983) Cellular mechanisms of insulin release. Effects of retinoids on rat islet cell-to-cell adhesion, reaggregation, and insulin release. Diabetes 32: 568–574
13. Halban PA, Wollheim CB, Blondel B, Meda P, Niesor EN, Mintz DH (1982) The possible importance of contact between pancreatic islet cells for the control of insulin release. Endocrinology 111: 86–94
14. Maes E, Pipeleers D (1984) Effects of glucose and 3′,5′-cyclic adenosine monophosphate upon reaggregation of single pancreatic B-cells. Endocrinology 114: 2205–2209
15. Lernmark A (1974) The preparation of, and studies on, free cell suspensions from mouse pancratic islets. Diabetologia 10:431–438
16. Pipeleers D, In't Veld P, Maes E, Van de Winkel M (1982) Glucose-induced insulin release depends on functional cooperation between islet cells. Proc Natl Acad Sci USA 79: 7322–7325
17. Salomon D, Meda P (1986) Heterogeneity and contact-dependent regulation of hormone secretion by individual B cells. Exp Cell Res 162: 507–520
18. Bosco D, Orci L, Meda P (1989) Homologous but not heterologous contact increases the functioning of individual secretory cells. Exp Cell Res 184: 72–80
19. Stagner JI (1991) Pulsatile secretion from the endocrine pancreas: metabolic, hormonal, and neural modulation. In: Samols E (ed) The endocrine pancreas. Raven, New York, pp 283–302
20. Bergsten P, Hellman B (1993) Glucose-induced amplitude regulation of pulsatile insulin secretion from individual pancreatic islets. Diabetes 42: 670–674
21. Kawai K, Ipp E, Orci L, Perrelet A, Unger RH (1982) Circulating somatostatin acts on the islets of Langerhans by way of a somatostatin-poor compartment. Science 218: 477–478
22. Samols E, Stagner JI, Ewart RBL, Marks V (1988) The order of islet cellular perfusion is B-A-D in the perfused rat pancreas. J Clin Invest 82: 1715–1721
23. Orci L, Unger RH (1975) Functional subdivision of islets of Langerhans and possible role of D cells. Lancet II: 1243–1244
24. Hellman B, Gylfe E, Grapengiesser E, Lund P-E, Berts A (1992) Cytoplasmic Ca^{2+} oscillations in pancreatic β-cells. Biochim Biophys Acta 1113: 295–305
25. Rorsman P, Trube G (1986) Calcium and delayed potassium currents in mouse pancreatic β-cells under voltage clamp conditions. J Physiol (Lond) 374: 531–550
26. Falke LC, Gillis KD, Pressel DM, Misler S (1989) "Perforated patch recording" allows long-term monitoring of metabolite-induced electrical activity and voltage-dependent Ca^{2+} currents in pancreatic B cells. FEBS Lett 251: 167–172
27. Meissner HP (1976) Electrophysiological evidence for coupling between pancreatic B cells of pancreatic islets. Nature 262: 502–504
28. Eddlestone GT, Gonçalves A, Bangham JA, Rojas E (1984) Electrical coupling between cells in islets of Langerhans from mouse. J Membr Biol 77: 1–14
29. Meda P, Atwater I, Gonçalves A, Bangham A, Orci L, Rojas E (1984) The topography of electrical synchrony among B-cells in the mouse islets of Langerhans. Q J Exp Physiol 69: 719–735
30. Valdeolmillos M, Santos RM, Contreras D, Soria B, Rosario LM (1989) Glucose-induced oscillations of intracellular Ca^{2+} concentration resembling bursting electrical activity in single mouse islets of Langerhans. FEBS Lett 259: 19–23
31. Valdeolmillos M, Nadal A, Soria B, Garcia-Sancho J (1993) Fluorescence digital image analysis of glucose-induced $[Ca^{2+}]_i$ oscillations in mouse pancreatic islets of Langerhans. Diabetes 42: 1210–1214
32. Pérez-Armendariz E, Atwater I, Rojas E (1985) Glucose-induced oscillatory changes in extracellular ionized potassium concentration in mouse islets of Langerhans. Biophys J 48: 741–749
33. Kumar NM, Gilula NB (1992) Molecular biology and genetics of gap junction channels. Semin Cell Biol 3: 3–16

34. Willecke K, Hennemann H, Dahl E, Jungbluth S, Heynkes R (1991) The diversity of connexin genes encoding gap junctional proteins. Eur J Cell Biol 56: 1–7
35. Meda P, Pepper M, Traub O, Willecke K, Gros D, Beyer E, Nicholson B, Paul D, Orci L (1993) Differential expression of gap junction connexins in endocrine and exocrine glands. Endocrinology 133: 2371–2378
36. Meda P, Chanson M, Pepper M, Giordano E, Bosco D, Traub O, Willecke K, El Aoumari A, Gros D, Beyer E, Orci L, Spray DC (1991) In vivo modulation of connexin 43 gene expression and junctional coupling of pancreatic B-cells. Exp Cell Res 192: 469–480
37. Meda P, Perrelet A, Orci L (1979) Increase of gap junctions between pancreatic B-cells during stimulation of insulin secretion. J Cell Biol 82: 441–448
38. Meda P, Denef J-F, Perrelet A, Orci L (1980a) Nonrandom distribution of gap junctions between pancreatic B-cells. Am J Physiol 238: C114–C119
39. Meda P, Halban P, Perrelet A, Renold AE, Orci L (1980b) Gap junction development is correlated with insulin content in the pancreatic B-cell. Science 209: 1026–1028
40. Pérez-Armendariz M, Roy C, Spray DC, Bennett MVL (1991) Biophysical properties of gap junctions between freshly dispersed pairs of mouse pancreatic beta cells. Biophys J 59: 76–92
41. Meda P, Amherdt M, Perrelet A, Orci L (1981) Metabolic coupling between cultured pancreatic B-cells. Exp Cell Res 133: 421–430
42. Meda P, Kohen E, Kohen C, Rabinovitch A, Orci L (1982) Direct communication of homologous and heterologous endocrine islet cells in culture. J Cell Biol 92: 221–226
43. Kohen E, Kohen C, Thorell B, Mintz DH, Rabinovitch A (1979) Intercellular communication in pancreatic islet monolayer cultures: a microfluorometric study. Science 204: 862–865
44. Kohen E, Kohen C, Rabinovitch A (1983) Cell-to-cell communication in rat pancreatic islet monolayer cultures is modulated by agents affecting islet cell secretory activity. Diabetes 32: 95–98
45. Michaels RL, Sheridan JD (1981) Islets of Langerhans: dye coupling among immunocytochemically distinct cell types. Science 214: 801–803
46. Meda P, Michaels RL, Halban PA, Orci L, Sheridan JD (1983) In vivo modulation of gap junctions and dye coupling between B-cells of the intact pancreatic islet. Diabetes 32: 858–868
47. Bosco D, Meda P (1992) Actively synthetizing β-cells secrete preferentially during glucose stimulation. Endocrinology 129: 3157–3166
48. Philippe J, Giordano E, Gjinovci A, Meda P (1992) cAMP prevents the glucocorticoid-mediated inhibition of insulin gene expression in rodent islet cells. J Clin Invest 90: 2228–2233
49. Meda P, Bosco D, Chanson M, Giordano E, Vallar L, Wollheim C, Orci L (1990a) Rapid and reversible secretion changes during uncoupling of rat insulin-producing cells. J Clin Invest 86: 759–768
50. Bruzzone R, Meda P (1988) The gap junction: a channel for multiple functions? Eur J Clin Invest 18: 444–453
51. Vozzi C, Ullrich S, Charollais A, Philippe J, Orci L, Meda P (1995) Adequate connexin expression is required for proper insulin production. J Cell Biol 131: 1561–1572
52. Bosco D, Meda P, Thorens B, Malaisse WJ (1995) Heterogenous secretion of individual B-cells in response to D-glucose and to non-glucidic nutrient secretagogues. Am J Physiol 268: C611–C618
53. Soria B, Chanson M, Giordano E, Bosco D, Meda P (1991) Ion channels of glucose-responsive and -unresponsive B-cells. Diabetes 40: 1069–1078
54. Giordano E, Bosco D, Cirulli V, Meda P (1991) Repeated glucose stimulation reveals distinct and lasting secretion patterns of individual rat pancreatic B-cells. J Clin Invest 87: 2178–2185
55. Bosco D, Meda P (1994) Individual cell-to-cell contacts rapidly recruit pancreatic β-cells for glucose-induced insulin secretion. Acta Anat (Basel) 149: 148
56. Van Schravendijk CFH, Kiekens R, Pipeleers DG (1992) Pancreatic β cell heterogeneity in glucose-induced insulin secretion. J Biol Chem 267: 21344–21348
57. Giordano E, Cirulli V, Bosco D, Rouiller D, Halban P, Meda P (1993) B-cell size influences glucose-stimulated insulin secretion. Am J Physiol 265: C358–C364
58. Kiekens R, In't Veld PA, Pipeleers DG (1992) Differences in glucose recognition by individual rat pancreatic B cells are associated with intercellular differences in glucose-induced biosynthetic activity. J Clin Invest 89: 117–125
59. Schuit FC, In't Veld PA, Pipeleers DG (1988) Glucose stimulates proinsulin biosynthesis by a dose-dependent recruitment of pancreatic beta cells. Proc Natl Acad Sci USA 85: 3865–3869

60. Johnston MF, Simons SA, Ramon F (1980) Interaction of anaesthetics with electrical synapses. Nature 286: 498–500

61. Musil LS, Goodenough DA (1991) Biochemical analysis of connexin43 intracellular transport, phosphorylation and assembly into gap junction plaques. J Cell Biol 115: 1357–1374

62. Stefan Y, Meda P, Neufeld M, Orci L (1987) Stimulation of insulin secretion reveals heterogeneity of pancreatic B-cells in vivo. J Clin Invest 80: 175–183

63. Hooper ML, Subak-Sharpe JH (1981) Metabolic co-operation between cells. Int Rev Cytol 69: 46–104

64. Gylfe E, Grapengiesser E, Hellman B (1991) Propagation of cytoplasmic Ca^{2+} oscillations in clusters of pancreatic β-cells exposed to glucose. Cell Calcium 12: 229–240

65. Longo EA, Tornheim K, Oeeney JT, Varnum BA, Tillotson D, Prentki M, Corkey BE (1991) Oscillations in cytosolic free Ca^{2+}, oxygen consumption, and insulin secretion in glucose-stimulated rat pancreatic islets. J Biol Chem 266: 9314–9319

66. Santos RM, Rosario LM, Nadal A, Garcia-Sancho J, Soria B, Valdeomillos M (1991) Widespread synchronous $[Ca^{2+}]_i$ oscillations due to bursting electrical activity in single pancreatic islets. Pflugers Arch 418: 417–422

67. Sorenson RL, Parsons JA (1985) Insulin secretion in mammosomatotropic tumor-bearing and pregnant rats. A role for lactogens. Diabetes 34: 338–341

68. Atwater I, Rosario L, Rojas E (1983) Properties of the Ca-activated K^+ channel in pancreatic B-cells. Cell Calcium 4: 451–461

69. Lawrence TS, Beers WH, Gilula NB (1978) Transmission of hormonal stimulation by cell-to-cell communication. Nature 272: 501–506

70. Murray SA, Fletcher WH (1984) Hormone-induced intercellular signal transfer dissociates cyclic AMP-dependent protein kinase. J Cell Biol 98: 1710–1719

71. Stagg RB, Fletcher WH (1990) The hormone-induced regulation of contact-dependent cell-cell communication by phosphorylation. Endocr Rev 11: 302–325

72. Rasmussen H (1991) Disordered cell communication as the basis of human disease: Implications for 21st-century medicine. In: Hardy MA, Kinne RKH (eds) Biology and medicine into the 21st century. Karger, Basel, pp 33–68

73. Unger RH, Foster DW (1992) Diabetes Mellitus. In: Wilson JD, Foster DW (eds) Williams textbook of endocrinology 8th edn. Saunders, Philadelphia, pp 1255–1333

74. Cerasi E, Luff R, Efendic S (1971) Decreased sensitivity of the pancreatic beta cells to glucose in pre-diabetic and diabetic subjects. A glucose dose-response study. Diabetes 21: 224–234

Expression of the Insulin Gene and Its Regulation

J. Philippe

Introduction

Maintenance of the blood glucose concentration within a narrow range requires a delicate balance between glucose production and utilization; this can only be realized through the interplay of two major hormones, insulin and glucagon. Insulin is synthesized by the β-cells of the pancreatic islets of Langerhans whereas glucagon is produced by the α-cells. Biosynthesis and secretion of these two hormones must be tightly regulated to match fuel production and delivery to metabolic demands. Insulin is secreted in response to nutrients (glucose and amino acids) to promote energy storage in target organs (liver, muscle and adipose tissue); glucagon secretion is then inhibited. In the fasted state, insulin secretion falls and glucagon release is stimulated; this results in an activation of glycogenolysis and gluconeogenesis and, despite the lack of exogenous glucose, in the maintenance of a stable blood glucose concentration.

In order to adapt to the absorptive and postabsorptive states, the α- and β-cells of the islets of Langerhans need to sense the ambient glucose concentration and to modulate biosynthesis and secretion of their respective hormonal products.

In this review we will examine how the insulin gene is specifically expressed in the β-cell and how it is regulated by glucose. These two processes, which could at first appear as completely different and independent are in fact closely linked, indicating that production of insulin and its regulation by glucose, the two major reasons for the existence of a β-cell, are determined by the same molecular mechanisms.

Glucose is the major but not the sole regulator of insulin gene expression. Second messengers, particularly cyclic adenosine monophosphate (cAMP), have been involved in the regulation of the insulin gene and we will review the relevant data.

Structure of the Insulin Gene

The structure of the insulin gene and/or cDNA sequences have been characterized in humans, Old World monkeys *Macaca fasciculari*, dogs, rats, mice, Syrian hamsters, guinea pigs, chickens, carp, angler fish, salmon and hagfish [1]. The nucleotide and the deduced amino acid sequences of the insulin gene are highly conserved throughout evolution [1]. An interesting exception is guinea pig insulin which has accumulated mutations in the A and B chains at the same frequency as in the signal sequence, with the C-peptide resulting in a decreased potency compared to insulins from other species. The lower metabolic effects of insulin in the guinea pig may, however, be

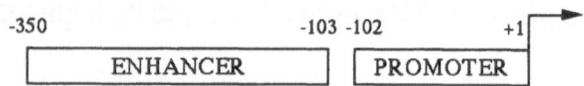

Fig. 1. The rat insulin I gene 5′-flanking region can be divided into a proximal promoter and a distal enhancer which both confer β-cell specificity to insulin gene expression

compensated by a glucagon molecule characterized by the substitution of five amino acids at the carboxy-terminal end and which displays a reduced binding affinity to its own receptor compared to other mammalian glucagons [2,3].

The insulin gene exists as a single copy gene in most species except rodents (rats and mice) where two nonallelic and functional insulin genes are expressed. The human insulin gene (Fig. 1) contains three exons separated by two introns [1]. The introns are variable in length and sequence between species but the locations of both introns have been highly conserved; intron 1 is positioned in the 5′-untranslated region while intron 2 interrupts the coding sequence of the C-peptide.

In rodents, the two nonallelic insulin genes (insulin I and II) are more than 90% homologous in nucleotide sequence. Insulin I and II only differ by substitution of amino acids at positions B9 and B29. The structure of the rat and mouse insulin I gene differs, however, from that of other insulin genes; intron 2 is indeed lacking. In addition, the 5′-flanking sequences of the rat insulin I and II genes are homologous up to 540 bp upstream of the transcriptional start-site and then differ. These observations have led to the suggestion that the rat and mouse insulin I gene arose by RNA-mediated duplication of an incompletely processed insulin II RNA transcript initiated upstream from the normal initiation site [4].

The human insulin gene lies on chromosome 11, while the rat genes are located on chromosome 1 and the mouse genes on chromosomes 6 (gene I) and 7 (gene II) [1,5].

Developmental Regulation of the Insulin Gene

Most of the observations obtained on the developmental regulation of the insulin gene are derived from rat and mouse studies. The insulin gene is expressed as early as embryonic day (ED) 9.5 in the developing pancreas, and insulin messenger RNA levels subsequently increase until delivery. The relative levels of expression of the two insulin genes slightly differ during development [6].

Of major interest is the recent finding that the insulin gene is also expressed in extrapancreatic tissues (yolk sac, liver and brain) during development resulting in the production of insulin in these tissues [6,8]. Expression of the insulin gene in the yolk sac, liver and brain is limited to the ancestral II gene indicating that this gene, and not the insulin I gene, contains specific information that allows it to be transiently expressed in those tissues. Maximal expression is observed at ED 20 in the yolk sac of the rat and at EDs 13–16 in the liver. During that period, insulin mRNA represents 1/20 to 1/10 the abundance of total insulin mRNA in the adult pancreas indicating that relatively large amounts of insulin may be synthesized in the fetal liver. In the mouse, brain insulin II mRNA is detected early at ED 9.5; it then increases until ED 14.5 and disappears [6].

Nothing is known about the role of insulin in the yolk sac, liver and brain during development. Recent data, however, suggest that insulin is critical for normal brain development [9]. In transgenic mice, ablation of insulin-producing cells by selective diphtheria toxin gene expression results in severe brain malformation.

β-Cell Specific Expression of the Insulin Gene

The characteristics of the β-cell result from the specific expression of a limited number of genes whose products interact to establish a unique environment.

Among these genes, the most relevant is the insulin gene since insulin production is the raison d'être of the β-cell. How individual cells can selectively express certain transcription units present in the genome has been a major interest in molecular biology. Over the last few years we have learned that the interactions of transacting factors with cis-acting DNA elements present in gene promoters are critical to activate gene transcription. Transacting factors may even have a much wider role in the biology of the cell and participate in processes such as cell proliferation, differentiation and maintenance of the differentiated state. Characterization of the factors responsible for insulin gene expression should allow us to better understand how the β-cell functions.

The insulin gene 5'-flanking region represents sequences upstream of the transcription initiation site. It contains the necessary information for both β-cell specific and regulated expression in response to developmental and physiological stimuli. Most of the information on the role of the insulin gene promoter has been obtained from cell transfection experiments or transgenic mice using fusion genes consisting of the rat I, II or human insulin gene promoters linked to reporter genes.

The exact length of the 5'-flanking sequences of the insulin gene necessary for maximal basal and regulated transcriptional activity is unknown. Although studies in

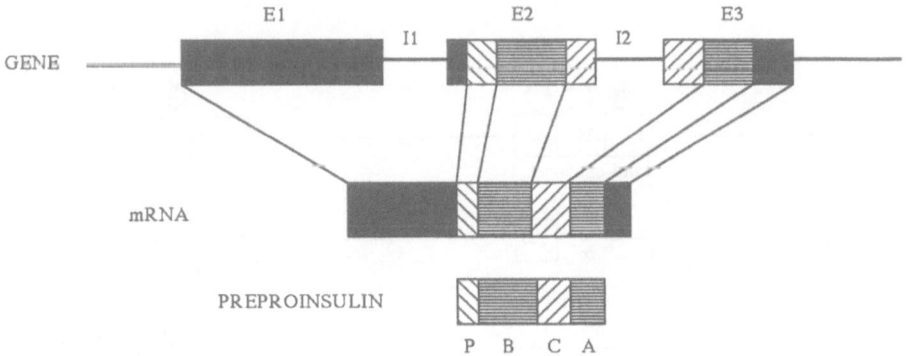

Fig. 2. Schematic representation of the human insulin gene and the encoded messenger RNA and preproprotein, preproinsulin. Correspondence between exons and functional domains of preproinsulin is indicated by *lines* between genes and mRNA and by specific boxes. *E* indicates exon and *I*, introns. ■ represent the 5'- and 3'-untranslated sequences; ▨, the signal peptide (*P*); ▦, the B and A chains; and ▧, the C-peptide (*C*)

transgenic mice with the human insulin gene have suggested that 4 kb of the 5'-flanking region are necessary for maximal activity [10,11], most work performed with transfected insulin-producing cells has focused on 400 bp.

We will first examine the rat insulin I gene, since sequences within its 5'-flank have been best characterized. We will then compare the structural organization of the *cis*-acting DNA elements between the two rat genes and the human gene.

Experiments already published 10 years ago suggested that the first 400 bp of the gene 5'-flanking sequence could be divided into a proximal promoter element up to–100 bp relative to the transcriptional start site and a distal enhancer (Fig. 2) [12,13]. Both elements have been shown to confer β-cell specific expression to the rat insulin I gene in cell transfection studies.

The distal enhancer (−346 bp to −103 bp) linked to a heterologous promoter is capable of directing expression to the β-cells but also, to a lesser extent, to the brain, suggesting that it is a major element, although not sufficient. Multiple nuclear protein binding sites have later been characterized within these large elements. Among these binding sites two different motifs now appear to be critical for β-cell specific transcription: the E motif whose consensus sequence is CANNTG, and the A motif represented by the sequence TAAT.

Of interest, these two motifs are present in multiples copies within the 5'-flanking region of the insulin I gene (Fig. 3). The E motif is present twice in the distal enhancer, whereas the A site appears both in the distal enhancer and in the proximal promoter. In the distal enhancer, the A and E motifs form a functional mini-enhancer unit which contains two A (A3 and A4) and one E motif (E2). As single motifs, these sequences exhibit weak or absent enhancer activity; even multiple copies of the E2 motif linked to the rat insulin I gene or to a heterologous promoter transactivate weakly. By contrast, when E2 and A3/A4 retain their natural context, they synergistically induce transcription [14–16].

More proximally, the A (A1) and E (E1) motifs are slightly more distant from each other compared to A3/A4 and E2 and it is unclear whether these two motifs can exhibit synergistic transcriptional activity.

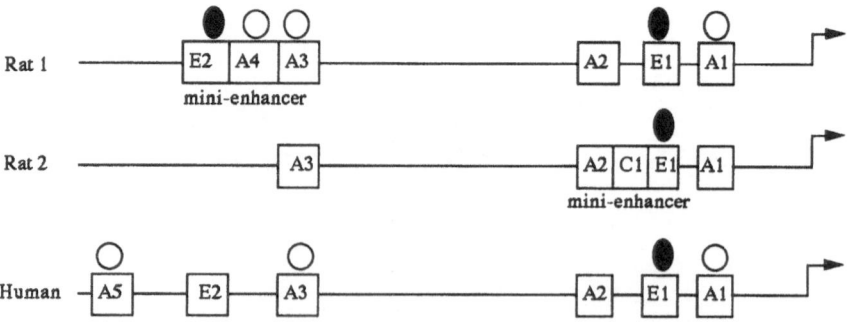

Fig. 3. *Cis*-acting DNA elements of the rat I, II and human genes which are considered to play a role in the β-cell specific expression. (The nomenclature is derived from [51].) *White circles,* IPF-1; *black circles,* IEF-1; *circles* are indicated only when interactions between these proteins and the respective binding sites have been demonstrated

E1 and E2 correspond to the palindromic sequence CATCTG, which serves as recognition motifs for transcription factors characterized by a basic DNA-binding domain and a helix-loop-helix motif involved in both binding and dimerization. These factors bind DNA as homo- or heterodimers and have been shown to participate through their specific interaction with E boxes in the cell-specific expression of immunoglobulin and muscle-specific genes, as well as pituitary genes [17–19].

E1 and E2 bind the same protein complex, IEF, which consists of a ubiquitous helix-loop-helix protein similar or identical to E12/E47 [20], heterodimerized with a factor present in both β- and α-cells [21–23]. In β-cells, IEF-1 may be involved in the control of other transiently or permanently expressed genes such as the gastrin and secretin genes [24,25], whereas in α-cells IEF-1 participates in the control of glucagon gene expression [26].

Whereas the E motifs bind islet-specific complexes, the A binding sites interact with cell-specific proteins. Multiple candidates have been isolated and it is proposed that these bind the A site and transactivate the insulin gene promoter. The most promising candidate is now clearly IPE-1, a mouse homeo box containing DNA-binding protein which is expressed in most or all β-cells and in about 20% of the somatostatin-producing δ-cells [27,29].

IPF-1 binds to at least three A boxes of the rat insulin I gene (A1, A3 and A4) [27,29] and synergizes with one component of the E-box-binding factors IEF-1 (E47) to direct high level expression of the insulin enhancer [28]. STF-1 contains an N-terminal transactivation domain that is critical for activating the insulin gene. The homeo domain is not only important for DNA binding but also appears to be required for synergism with E 47 [28]. It is unknown, however, how the two complexes synergize to activate transcription, but it does not appear to involve cooperative DNA binding.

Cell transfection studies thus indicate that IPF-1 plays a major role, together with additional factors such as IEF-1, in insulin gene expression. IPF-1 has a wider role in development, however. In mouse embryos, IPF 1 expression is restricted to the developing pancreatic anlagen and is initiated when the foregut endoderm is committed to a pancreatic fate. In mice homozygous for a targeted mutation in the IPF-1 gene the pancreas gland is completely lacking, whereas the gastrointestinal tract and the other abdominal organs are normal [32]. These data thus suggest that IPF-1 plays a very important function in the commitment of early foregut endodermal cells to pancreatic development.

IPF-1 has also been proposed to be involved in somatostatin gene expression in islet cells [28,29]; IPF-1 indeed binds to two tissue-specific elements, TSE-I and TSE-II and activates an intact somatostatin promoter; the fact that IPF-1 is only present in 20% of somatostatin-producing cells indicates, however, that it is not required for the maintenance of somatostatin gene expression in the adult. Whether IPF-1 is necessary for somatostatin gene regulation or for δ-cell differentiation during development will need additional studies.

Other DNA control elements have been described in the 5'-flanking region of the rat insulin I gene, but their importance in tissue-specificity is not clearly established.

Comparison between the structural organization of the rat insulin I, II and human genes reveals clear differences, although A and E binding motifs are found in all three genes. The functional equivalent of the rat insulin I mini-enhancer in the rat insulin II gene is represented by the 30 bp RIPE3 element (−86 to −126 bp) that contains an E box proximally and two upstream binding sites, C1 and A2. RIPE3 may be sufficient to confer β-cell specific expression in insulinoma cells and in transgenic mice [33,34]. Although the E box binds IEF-1, the factors interacting with C1 and A2 have not been characterized. In addition, these factors may be different from IPF-1, although synergistic activation of the rat insulin II gene promoter has been observed in co-transfection experiments with the IPF-1 and E47 cDNAs [28]. It remains to be demonstrated whether IPF-1 binds A2 to activate insulin II gene transcription synergistically with IEF-1 and/or whether it binds any additional A sites.

IPE-1 also appears to be involved in the regulation of the human insulin gene as it binds to the A1, A3 and A4 elements and activates transcription. Binding affinity is different for the three sites, being higher for A3 than for A1, and lower for A4.

A functional equivalent of the rat insulin I mini-enhancer has not been identified in the human gene. The human gene contains only one E box, E1, which is found just upstream of the A1 site. It is thus possible that the A1/E1 binding sites, with or without A2, represent a functional unit allowing IPF-1 and IEF-1 to synergize to activate transcription.

Insulin Gene Regulation by Glucose

Glucose is the major regulator of insulin secretion and biosynthesis. Glucose acts at multiple steps to increase insulin biosynthesis [35–37]. Acute stimulation is dependent on translation initiation and peptide chain elongation. On-going biosynthesis is unaffected at this stage by inhibitors of mRNA formation [35]. Glucose also acts by modulating insulin mRNA half-life and gene transcription [36,37]. The mechanism by

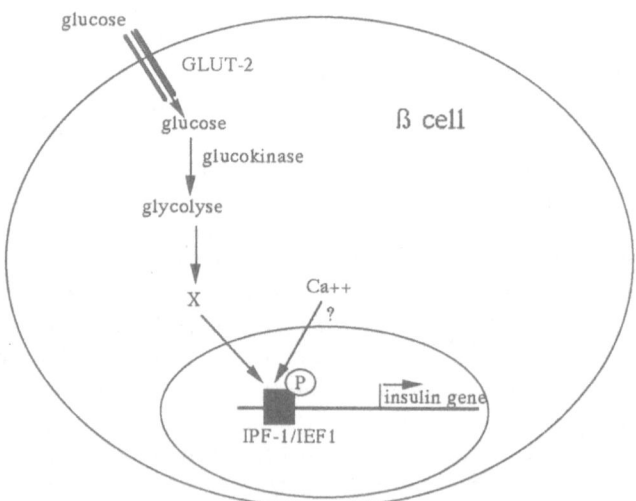

Fig. 4. Activation of insulin gene transcription by glucose. X indicates an unknown factor responsible for transmitting the glucose signal to transcription factors, possibly IPF-1/IEF-1 by phosphorylation (P)

which glucose exerts its effect on transcription has been the subject of very informative studies performed during the last 5 years. The 5′-flanking DNA of the insulin I gene was shown to confer glucose responsiveness to a reporter gene when expressed in β-cells [38]. A glucose-responsive element was then localized by German et al.; this element corresponds to the mini-enhancer. Interestingly, the binding activity of the mini-enhancer A elements most likely represented by IPF-1 is correlated with glucose concentrations (Fig. 4) [39,40]. In low glucose, binding activity decreases rapidly (within 6 h) to low levels, whereas it recovers rapidly within 15 min in high glucose; the effect is inhibited by mannoheptulose, indicating that it is dependent on glucose metabolism. In addition, phosphatase inhibitors are capable of preventing the fall in binding activity [40]. The role of the E2 box in the glucose effect is not entirely clear. Using fetal islet cells, German found a critical role for E2 and demonstrated a glucose-stimulated increase in IEF-1 binding [38,41], while in adult islets E2 was not necessary for the glucose response [39].

IPE-1 also appears to be involved in the regulation of the human insulin gene by glucose through its binding to the A3 box [39,42]. Whether glucose also acts through the A1 and A4 boxes is still unknown. Glucose regulation of the rat insulin II gene occurs through the RIPE3 element. Both the E1 box and the upstream C1/A2 sequences are necessary for a maximal glucose effect in HIT-T15 cells, although only the upstream binding affinity increases with high glucose [43]. Interestingly, the increase in insulin gene transcription by glucose is accompanied by a concomitant decrease in c-jun mRNA levels. This observation, combined with the fact that the jun proteins can inhibit insulin gene transcription through interaction with IEF-1, suggest that E1 may also be a target of glucose regulation [43].

Since there are multiple IPF-1 and IEF-1 binding sites within the insulin I and human gene promoters, glucose responsiveness may be mediated by multiple control elements. This has in fact been proposed for the rat insulin I gene in as much as no single mutation of the promoter was able to entirely remove its ability to respond to glucose [41]. It is thus possible that glucose regulates insulin gene transcription through multiple DNA control elements which remain to be clearly localized and defined within the respective insulin gene promoters.

How changes in glucose concentrations are transmitted to the nucleus are for the most part unknown (Fig. 4). Cells sense glucose concentrations by the levels of the products of glucose metabolism. Glucokinase acts as a glucose sensor and its activity is critical for appropriate insulin biosynthesis. It has been proposed that intermediates of glucose metabolism and calcium transmit the glucose signal to the nucleus [38]. Recent data indicate that eventually glucose results in the phosphorylation of transacting factors and affects the transcriptional machinery by unknown mechanisms. The kinase(s) responsible for IPF-1 phosphorylation have not been identified, but protein kinase A and C are unlikely to be involved [40].

Insulin Gene Regulation by cAMP

Activation of the cAMP second messenger pathway by the interactions of membrane receptors with their respective ligands results in a cascade of biochemical reactions leading to the regulation of specific genes. Insulin gene transcription and insulin

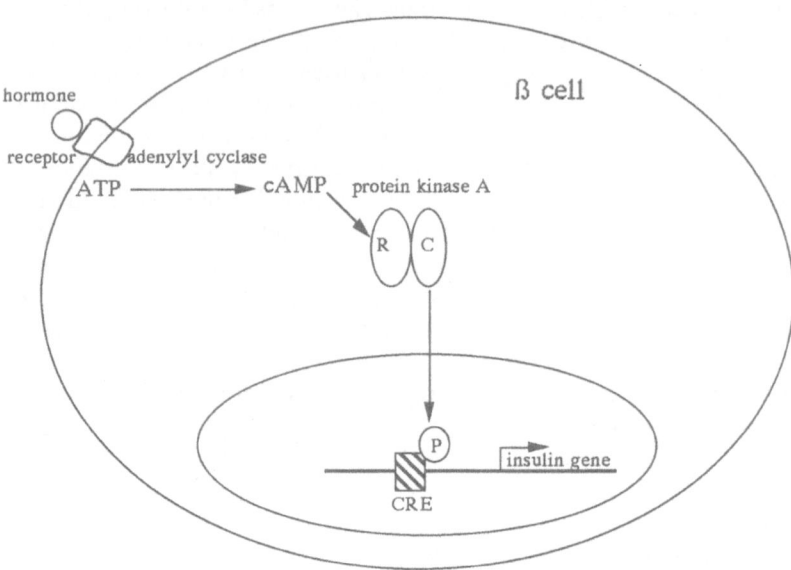

Fig. 5. Activation of insulin gene transcription by cAMP. *R* and *C* indicate the regulatory and catalytic units, respectively, of protein kinase A; *CRE*, the cAMP-responsive DNA element, and *P* a transcription factor phosphorylated in response to the activation of protein kinase A

mRNA half-life have been shown to be acutely affected by cAMP analogues in both isolated islets and transformed cell lines [36,37]. A cAMP-responsive element (CRE) localized between nucleotides −177 and −184 has been characterized within the rat insulin I gene [44] (Fig. 5).

The CRE of the insulin gene exhibit specific properties compared to the CRE of the glucagon and somatostatin genes [45]. The latter two CREs bind the nuclear CRE binding protein (CREB) which, on phosphorylation on serine 133 by protein kinase A, results in an activation of transcription [46]. Although recombinant CREB can bind the insulin CRE effectively, cellular CREB binds very poorly, indicating that post-translational modifications may be responsible for this discrepancy. In fact, protein complexes which bind the insulin CRE in vitro are not characterized. Functionally, the insulin CRE confers basal activity but relatively weak cAMP responsiveness compared to the glucagon and somatostatin genes' CREs. In addition it is not activated by membrane depolarization and calcium influx, a property of the two other CREs [45]. The characteristics of the insulin CRE distinguish it functionally from more classical CREs and are likely to be physiologically important for the regulation of insulin gene transcription. Sequence comparison for the presence of the CRE motif within the insulin gene promoter of different species reveals similar elements (TGACGTCC) within the human, mouse and rat insulin II genes. Furthermore, the corresponding sequence of the human insulin gene promoter has been shown to confer cAMP inducibility [47].

Different hormones which affect intra-cellular cAMP levels may regulate insulin gene expression [48–50]. Glucagon-like peptide I is a powerful and physiologically important insulin secretagogue which also stimulates insulin gene transcription through changes in cAMP levels. Galanin, epinephrine and somatostatin have also been proposed to affect insulin gene expression through the cAMP second messenger pathway.

Conclusions

Recent data indicate that the mechanisms involved in both β-cell specific expression of the insulin gene and its regulation by glucose converge on one or two transcription factors, IPF-1 and IEF-1. IPF-1 appears to be involved not only in these critical β-cell functions but also in the formation of the pancreas in the early stages of development, illustrating the multiple roles of transcription factors.

Although we have witnessed considerable progress with the isolation of IPF-1, much remains to be learned on insulin gene expression and islet cell differentiation. For instance, the presence of IPF-1 in 20% of somatostatin-producing cells and of IEF-1 in phenotypically different islet cells suggests that additional factors may be required for the cell-specific expression of the insulin gene.

The next decade promises to bring major insights into the latter processes.

References

1. Steiner DF, Chan SJ, Welsh MJ, Kwok SCM (1985) Structure and evolution of the insulin gene. Annu Rev Genet 19: 463–484
2. Chan SJ, Episkopon V, Zeitlin S, Karathanasis SK, McKrell A, Steiner DF, Efstratiadis A (1984) Guinea pig preproinsulin gene: an evolutionary compromise? Proc Natl Acad Sci USA 81: 5046–5050
3. Seino S, Welsh M, Bell GI, Chan SJ, Steiner DF (1986) Mutations in the guinea pig preproglucagon gene are restricted to a specific portion of the prohormone sequence. FEBS Lett 203: 25–30
4. Soares MB, Schon E, Henderson A, Karathanasis SK, Cate R, Zeitlin S, Chirgwin J, Efstratiadis A (1985) RNA-mediated gene duplication: the rat preproinsulin I gene is a functional retroposon. Mol Cell Biol 5: 2090–2103
5. Owerbach D, Bell GI, Rutter WJ, Shows TB (1980) The insulin gene is located on chromosome 11 in humans. Nature 286: 82 84
6. Deltour L, Leduque P, Blume N, Madsen O, Dubois P, Jami J, Bucchini D (1993) Differential expression of the nonallelic proinsulin genes in the developing mouse embryo. Proc Natl Acad Sci USA 90: 527–531
7. Muglia L, Locker J (1984) Extrapancreatic insulin gene expression in the fetal rat. Proc Natl Acad Sci USA 81: 3635–3639
8. Giddings SJ, Carnaghi LR (1990) Selective expression and developmental regulation of the ancestral rat insulin II gene in fetal liver. Mol Endocrinol 4: 1363–1369
9. Herrera PL, Huarte J, Zufferey R, Nichols A, Mermillod B, Philippe J, Muniesa P, Sanvito F, Orci L, Vassali JD (1994) Ablation of islet endocrine cells by targeted expression of hormone-promoter-driven toxigenes. Proc Natl Acad Sci USA 91: 12999–13003
10. Selden RF, Skoskiewics MJ, Burke Howie K, Russel PS, Goodman HS (1986) Regulation of human insulin gene expression in transgenic mice. Nature 321: 525–528
11. Bucchini D, Ripoche MA, Stinnakre MG, Desbois P, Lores P, Monthioux E, Absil J, Lepesant JA, Pictet R, Jami J (1986) Pancreatic expression of human insulin in transgenic mice. Proc Natl Acad Sci USA 83: 2511–2515
12. Walker MD, Edlund T, Boulet AM, Rutter WJ (1983) Cell-specific expression controlled by the 5′-flanking region of insulin and chymotrypsin genes. Nature 306: 557–561

13. Edlund T, Walker MD, Barr PJ, Rutter WJ (1985) Cell-specific expression of the rat insulin gene: evidence for role of two distinct 5'-flanking elements. Science 230: 912–916

14. Dandroy-Dron F, Monthioux E, Jami J, Bucchini D (1991) Regulatory regions of rat insulin I gene necessary for expression in transgenic mice. Nucleic Acids Res 19: 4925–4930

15. German MS, Wang J, Chadwick RB, Rutter WJ (1992) Synergistic activation of the insulin gene by a LIM-homeo domain protein and a basic helix-loop-helix protein: building a functional insulin mini-enhancer complex. Genes Dev 6: 2165–2176

16. German MS, Moss LG, Wang J, Rutter WJ (1992) The insulin and islet amyloid polypeptide genes contain similar cell-specific promoter elements that bind identical beta cell nuclear complexes. Mol Cell Biol 12: 1777–1788

17. Kadesch T (1992) Helix-loop-helix proteins in the regulation of immunoglobulin gene transcription. Immunol Today 13: 31–37

18. Wright WE (1992) Muscle basic helix-loop-helix proteins and the regulation of myogenesis. Curr Opin Genet Dev 2: 243–248

19. Therrien M, Drouin J (1993) Cell-specific helix-loop-helix factor required for pituitary expression of the proopiomelanocortin gene. Mol Cell Biol 13: 2342–2353

20. Murre C, Schonleber McCow P, Baltimore D (1989) A new DNA binding and dimerization motif in immunoglobulin enhancer binding, daughterless, MyoD, and myc proteins. Cell 56: 777–783

21. Aronheim A, Ohlsson H, Park CW, Edlund T, Walker MD (1991) Distribution and characterization of helix-loop-helix enhancer-binding proteins from pancreatic beta cells and lymphocytes. Nucleic Acids Res 19: 3893–3899

22. Park CW, Walker MD (1992) Subunit structure of cell-specific E box-binding proteins analyzed by quantitation of electrophoretic mobility shift. J Biol Chem 267: 15642–15649

23. Ohlsson H, Thor S, Edlund T (1991) Novel insulin promoter- and enhancer-binding proteins that discriminate between alpha and beta cells. Mol Endocrinol 5: 897–904

24. Wang TC, Brand SJ (1990) Islet cell-specific regulatory domain in the gastrin promoter contains adjacent positive and negative elements. J Biol Chem 265: 8908–8914

25. Wheeler MB, Nishitani J, Buchan AMJ, Kopin AS, Chey WY, Chang TM, Leiter AB (1992) Identification of a transcriptional enhancer important for entero-endocrine and pancreatic islet cell-specific expression of the secretin gene. Mol Cell Biol 12: 3531–3539

26. Cordier-Bussat M, Morel C, Philippe J (1995) Homologous DNA sequences and cellular factors are implicated in the control of glucagon and insulin gene expression. Mol Cell Biol 15: 3904–3916

27. Ohlsson H, Karlsson K, Edlund T (1993) IPF1, a homeo-domain-containing transactivator of the insulin gene. EMBO J 12: 4251–4259

28. Leonard J, Peers B, Johnson T, Ferreri K, Lee S, Montminy M (1993) Characterization of somatostatin transactivating factor-1, a novel homeobox factor which stimulates somatostatin expression in pancreatic islet cells. Mol Endocrinol 7: 1275–1283

29. Peers B, Leonard J, Sharma S, Teitelman G, Montminy MR (1994) Insulin expression in pancreatic islet cells relies on cooperative interactions. Mol Endocrinol 8: 1798–1806

30. Miller C, McGehee R, Habener J (1994) IDX-1: a new homeodomain transcription factor expressed in rat pancreatic islets and duodenum that transactivates the somatostatin gene. EMBO J 13: 1145–1156

31. Peshavaria M, Gamer L, Henderson E, Teitelman G, Wright C, Stein R (1994) XIHbox 8, an endoderm-specific Xenopus homeo domain protein, is closely related to a mammalian insulin gene transcription factor. Mol Endocrinol 8: 806–816

32. Jonsson J, Carlsson L, Edlund T, Edlund H (1994) Insulin-promoter-factor 1 is required for pancreas development in mice. Nature 371: 606–609

33. Shieh SY, Tsai MJ (1991) Cell-specific and ubiquitous factors are responsible for the enhancer activity of the rat insulin II gene. J Biol Chem 266: 16708–16714

34. Stellrecht CM, Finegold MJ, DeMayo MJ, Tsai MJ (1993) Tissue-specific activity of the rat insulin II gene enhancer, RIPE3, in transgenic mice. Proceedings of the 75th Annual Meeting of The Endocrine Society, Las Vegas, pp 373 (abstract)

35. Permutt MA (1974) Insulin biosynthesis IV Effect of glucose on initiation and elongation rates in isolated rat pancreatic islets J Biol Chem 249: 2738–2742

36. Nielsen DA, Welsh M, Casadaban MJ, Steiner DF (1985) Control of insulin gene expression in pancreatic beta cells and in an insulin-producing cell line, RIN 5F cells. Effects of glucose and cAMP on the transcription of insulin mRNA. J Biol Chem 260: 13585–13589

37. Welsh M, Nielsen DA, MacKrell AJ, Steiner DF (1985) Control of insulin gene expression in pancreatic beta cells and in an insulin-producing cell line, RIN 5F cells. II. Regulation of insulin mRNA stability. J Biol Chem 260: 13590–13594

38. German MS, Moss LG, Rutter WJ (1990) Regulation of insulin gene expression by glucose and calcium in transfected primary islet cultures. J Biol Chem 265: 22063–22066

39. Melloul D, Ben-Neriah Y, Cerasi E (1993) Glucose modulates the binding of an islet-specific factor to a conserved sequence within the rat I and the human insulin promoters. Proc Natl Acad Sci USA 90: 3865–3869

40. MacFarlane W, Read M, Gilligan M, Bujalska I, Docherty K (1994) Glucose modulates the binding activity of the beta-cell transcription factor IUF1 in a phosphorylation-dependent manner. Biochem J 303: 625–631

41. German MS, Wang J (1994) The insulin gene contains multiple transcriptional elements that respond to glucose. Mol Cell Biol 14: 4067–4075

42. Petersen H, Serup P, Leonard J, Michelsen BK, Madsen OD (1994) Transcriptional regulation of the human insulin gene is dependent on the homeodomain protein STF1/IPE1 acting through the CT boxes. Proc Natl Acad Sci USA 91: 10465–10469

43. Sharma A, Stein R (1994) Glucose-induced transcription of the insulin gene is mediated by factors required for beta-cell-type-specific expression. Mol Cell Biol 14: 871–879

44. Philippe J, Missotten M (1990) Functional characterization of a cAMP-responsive element of the rat insulin I gene. J Biol Chem 265: 1465–1469

45. Oetjen E, Diedrich T, Eggers A, Eckert B, Knepel W (1994) Distinct properties of the cAMP-responsive element of the rat insulin I gene. J Biol Chem 269: 27036–27044

46. Gonzalez GA, Montminy MR (1989) Cyclic AMP stimulates somatostatin gene transcription by phosphorylation of CREB at serine 133. Cell 59: 675–680

47. Inagaki N, Maekawa T, Sudo T, Ishii T, Seino S, Imura H (1992) C-jun represses the human insulin promoter activity that depends on multiple cAMP response elements. Proc Natl Acad Sci USA 89: 1045–1049

48. Fehmann HC, Habener JF (1992) Galanin inhibits proinsulin gene expression stimulated by the insulinotropic hormone glucagon-like peptide I (7-37) in mouse insulinoma BTC-1 cells. Endocrinoogy 130: 159–166

49. Zhang IJ, Redmon JB, Andresen JM, Robertson RP (1991) Somatostatin and epinephrine decrease insulin messenger ribonucleic acid in HIT cells through a pertussis toxin-sensitive mechanism. Endocrinology 129: 2409–2414

50. Philippe J (1993) Somatostatin inhibits insulin gene expression through a posttranscriptional mechanism in a hamster islet cell line. Diabetes 42: 244–249

51. German M, Ashcroft S, Docherty K et al. (1995) The insulin gene promoter: a simplified nomenclature. Diabetes 44: 1002–1009

Implications of the Glucokinase Glucose Sensor Paradigm for Pancreatic β-Cell Function

F.M. Matschinsky

Introduction

The β-cell glucokinase glucose sensor component of the feedback loops which regulate blood sugar is the primary topic of this chapter. This special aspect of the system should be viewed within the broader context of accepted tenets of blood sugar control: Normal pancreatic β-cell function is central to glucose homeostasis; precisely regulated glucose metabolism of the β-cells is required for the glucostat function of these cells; the glucokinase glucose sensor paradigm is the key to understanding β-cell glucose metabolism and stimulus secretion coupling; the insulin receptor and its signalling pathways and hormonal as well as autonomic counter regulatory processes complete the multicomponent feedback system that maintains glucose homeostasis.

Definition and Implications of the Glucokinase Glucose Sensor Concept

Glucose stimulation of insulin secretion is controlled by β-cell glucokinase [1–3]. The activity and glucose K_m of glucokinase and the cytosolic levels of glucose and ATP determine the rate of production of cytosolic and mitochondrial metabolic factors that couple metabolism of glucose to secretion. The most important among these are the reduced nicotinamide adenine dinucleotide (NADH) and the purine nucleotides ATP and ADP. Even a small reduction of glucokinase of 25%–50% by metabolic inhibitors or by genetic mutations is sufficient to lower glucose responsiveness of the β-cells and cause a diabetic phenotype. Normal glucokinase function is also a prerequisite for stimulation of insulin secretion by other physiological fuels (e.g., amino acids), hormones (e.g., enteroglucagon) and neurotransmitters (e.g., acetyl choline). The stimulation of most metabolic processes in β-cells, as manifested most strikingly by the 30%–50% enhancement of respiration by high glucose, does not require the presence of extracellular Ca^{2+} to be initiated [4–6]. This observation implies that glucokinase also governs mitochondrial equilibrium reactions and is a pacemaker for respiration and ATP production. Influx of extracellular Ca^{2+}, increased by glucose and other stimulatory fuel combinations, provides, however, the obligatory signal that causes insulin secretion. The independence from extracellular Ca^{2+} of the respiratory response is thus clearly distinguished from the dependence of the secretory response on extacellular Ca^{2+}. This difference is essential for a plausible hypothesis explaining glucose stimulation of secretion as a function of metabolic activation. In order to highlight this fundamental role of glucokinase in the β-cell glucostat function and to differentiate it from typical membrane-associated and soluble ligand-binding receptors, the term "glucokinase glucose sensor" was chosen [1–3] and has been adopted as useful by many investigators. The term appears to describe aptly the unique function of glucokinase in stimulus secretion coupling in pancreatic β-cells.

A Few Historical Remarks

Glucokinase was first demonstrated and quantitated in 1968 by Matschinsky and Ellerman in pancreatic islet tissue of the *ob/ob*-C57 black-6J mouse with a highly sensitive and specific fluorometric histochemical method [7]. The same publication also reported that the β-cells have a large capacity transport system for glucose that allows rapid equalization of intracellular glucose and of blood sugar. Those two observations provided the biochemical foundation for the glucokinase glucose sensor concept.

Earlier studies, beginning with a publication by Grodsky et al. in 1963 [8] and extended by Coore and Randle in 1964 [9] and Malaisse in 1968 [10], had concluded that glucose stimulation of insulin secretion required glucose phosphorylation and metabolism, but the precise enzymological nature of the glucose phosphorylation process could not be determined because of the limitation of the tissue preparations, shortcomings of the biochemical analytical methods and because specific metabolic inhibitors were not available. Ashcroft and Randle confirmed the presence of glucokinase in normal mouse pancreas in 1970 [11] and provided additional strong evidence for its essential role in glucose stimulation of insulin release. However, progress of science is dialectic. It is therefore not surprising that in the early 1970s many observations were made that did not seem to fit the metabolism or glucokinase glucose sensor concept and alternative hypotheses were sought to explain the large body of partly confusing information then available. Most striking of these new findings were the marked potentiation of pyruvate and tolbutamide actions by 2-desoxyglucose and mannoheptulose and the enhancement of glucose effects by galactose, all of which could not be readily incorporated in the glucokinase glucose sensor concept [12]. The state of knowledge was summarized by P. Randle [13] during a meeting in 1971 commemorating the 50th anniversary of the discovery of insulin: ". . . evidence currently available does not determine whether initiation of insulin secretion by glucose . . . involves a direct effect of the sugar or of a metabolite". The presence in β-cells of a membrane-bound glucoreceptor of broad specificity offered a plausible and testable alternative explanation [14]. As a result of the intensive research effort of many laboratories, including that of the author, the membrane glucoreceptor hypothesis was unanimously rejected in the mid and late 1970s and the crucial general role of metabolism in intracellular signal generation during glucose stimulation of insulin secretion was fully established. In contrast, the more specific proposal of the glucokinase-glucose sensor concept has continued to meet very outspoken resistance [15]. However, during the last 15 years the evidence in its favour has become overwhelming owing to the contribution of many laboratories [1–4,16–18]. Following the cloning of the human glucokinase gene by Permutt and his associates [19,20] it was discovered by Froguel and collaborators [21] that the manifestation of diabetes in a large fraction of families with maturity-onset diabetes of the young (MODY) was linked to the glucokinase gene. Nearly 30 mutations of glucokinase have since been identified in MODY families from all over of the world. The effect of the mutation on the kinetics of the enzyme explains the diabetic phenotype [22,23]. Very surprisingly, linkage to the glucokinase locus has recently been observed in families with insulin-dependent diabetes mellitus (IDDM) [24]. This brief summary indicates that it required more that 25 years of intensive research by many laboratories to establish the glucokinase glucose sensor paradigm so pivotal for the role of β-cells in glucose homeostasis.

Physiologically Significant Characteristics of Glucokinase

The characteristics of glucokinase, the hexose phosphotransferase that is expressed in hepatocytes, in pancreatic β-cells and which may also be present in rare neuroendocrine cells of the brain and the intestinal mucosa have been the subject of several competent recent reviews [1–4,16–19,22,23]. For the purpose of this article, which has primarily a physiological/chemical orientation, a reiteration of only a few crucial biochemical features of the enzyme is sufficient. Glucokinase is a 50-kDa protein which catalyses the transfer of phosphate from Mg-ATP to the 6-hydroxyl group of hexoses. The substrate specificity is not as high as the enzyme's name might suggest. The enzyme phosphorylates the D-forms of glucose, mannose, fructose, 2-desoxyglucose, glucosamine, mannosamine and mannoheptulose, albeit with very different rates and affinities (Table 1). It is inhibited by N-acetyl-glucosamine, which is not a substrate. It is noteworthy that galactose is neither a substrate nor an inhibitor of human recombinant glucokinase. With glucose as a substrate there is inhibition of the enzyme by several substrate and nonsubstrate sugars. For example, mannoheptulose, glucosamine, 2-desoxyglucose and N-acetylglucosamine have been used to investigate the role of glucokinase in the pancreatic β-cell. Mannoheptulose is the most commonly employed inhibitor because of its high efficacy ($K_i \simeq 0.7$) and seemingly effective transport into the β-cells. It should, however, be realized that

Table 1. Sugar substrates and inhibitors of recombinant human islet glucokinase

Sugar	Activity (U/mg prot.)	K_m (mM)	^nH	K_i^a (mM)
D-glucose	134	5.3	1.41	na
D-mannose	175	24	1.38	10
D-fructose	260	424	1.31	nm
D-2 desoxyglucose	70	85	1.30	10
D-glucosamine[b]	9	4.4	1.04	1.7
D-mannosamine[b]	0.26	16	0.82	3.8
D-mannoheptulose[b]	0.24	7.5	1.01	0.78
D-N-acetylglucosamine	nd	nd	nd	0.14

1U, 1 μmol of product/min; na, not applicable; nd, not detectable; nm, not measured.
[a] The K_i was determined with racemic D-glucose and Mg-ATP as substrate. All kinetic constants were obtained spectrophotometrically at 30°C. A Q10 of 2 was used to calculate the activity for 37°C.
[b] The activities and K_m values for sugars that are substrates and inhibitors could be more apparent than real because the present studies were not extended to very high levels of substrate.

mannoheptulose is also a very powerful inhibitor of other hexose phosphotrans-ferases, for example hexokinase I (called brain hexokinase, $K_i \simeq 0.025\,\text{m}M$). This complicates the interpretation of the studies with mannoheptulose. The glucokinase K_m for racemic D-glucose is $4-8\,\text{m}M$. The enzyme prefers the α-anomer of its sub-strate as demonstrated for glucose and mannose. The enzyme shows sigmoidal de-pendency of activity on glucose concentration manifested by a Hill coefficient of about 1.5. Sigmoidal dependency of activity on the substrate level is observed with glucose, mannose, fructose and 2-desoxyglucose. This sigmoidicity is attributed to a mnemonical mechanism which envisions differential glucose binding to two forms of the enzyme that are not in equilibrium, or it has been explained by a slow transition model. The enzyme appears to show little influence by allosteric metabolite activators or modifiers. Long chain acyl coenzyme A (CoA) esters are, however, surprisingly powerful inhibitors (the K_i for stearyl-CoA is $1.5\,\mu M$). However, the enzyme is in-hibited by a regulatory protein in a fructose-6-phoshate (fructose-6-P) dependent manner. This indirect inhibition is abolished by fructose-1-P. The hexose-P depen-dent regulatory protein and acyl-CoA seem to bind to the same allosteric site of the enzyme. cDNAs for glucokinase and other hexose phosphotransferases have been obtained and the amino acid sequences have been established, including the primary structure of human glucokinase isoforms. The data suggest that a primordial glucokinase-like gene gave rise to a family of hexose phosphotransferases by gene duplication. A molecular model of human glucokinase based on the crystal structure of yeast hexokinase B has been proposed.

Biochemical Design Features of Fuel Metabolism of the Pancreatic β-Cells As Compared to Hepatocytes

The glucokinase glucose sensor concept arises from our knowledge of the specific biochemical design of β-cell fuel metabolism [1–4,16–19,22,23]. Its features are con-veniently highlighted by comparison with the well known biochemical design of liver cells (Fig. 1). Like hepatocytes, β-cells have a high capacity high K_m glucose trans-porter Glut-2. This permits rapid equalization of intracellular glucose with blood sugar levels and points to an intracellular sensing device for blood sugar. Like hepa-tocytes, β-cells have glucokinase or hexokinase IV. However, the β-cell glucokinase activity is almost one order of magnitude lower than that of liver and is controlled by ambient glucose rather than by insulin as is true in liver cells. Liver contains signifi-cant levels of the glucokinase regulatory protein, which mediates feedback inhibition of glucokinase by fructose-6-P. There is little evidence for an important physiological role of this regulatory protein in the β-cells. β-cells appear to have very low activity of the low K_m hexokinase, if any, similar to liver [25]. As an approximation for our calculations we use a V_{max} of $1/10$ and a K_m of $1/1000$ of the corresponding glucokinase parameters. We also assume that β-cell hexokinase is similar to hexokinase I of the brain and is inhibited by glucose-6-P, 6-P-gluconate and glucose-1, 6-P_2 and deinhibited by P_i. Much work has been done during the last 25 years to assess the role of glucose-6-phosphatase in β-cell glucose metabolism. It is now probable that a glucose-6-phosphatase with characteristics of the enzyme present in hepatocytes can-not be demonstrated in β-cells of normal, freshly isolated rat islets [26]. This does not

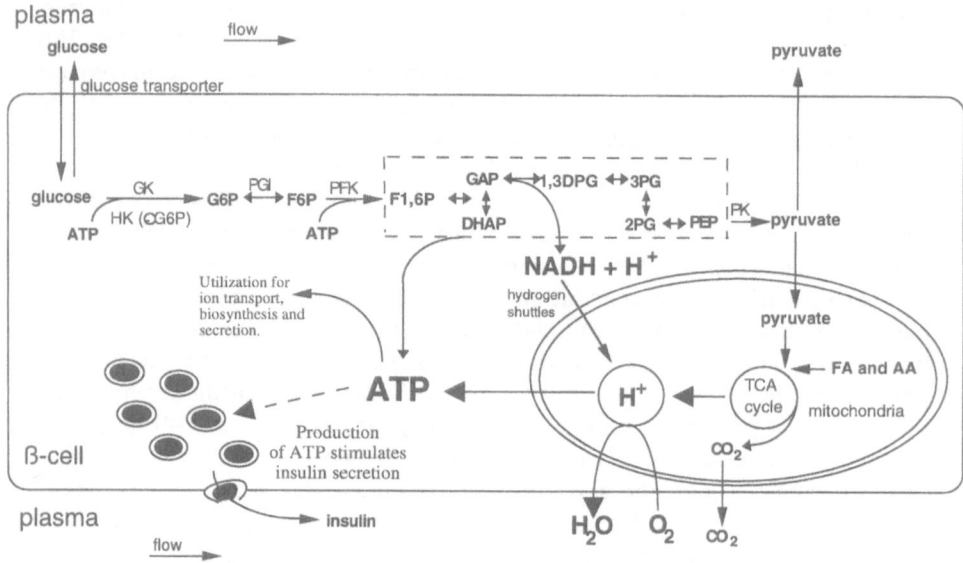

Fig. 1. *Conceptual model of glucose-induced insulin secretion in pancreatic β-cells.* The model was constructed from the perspective that the phosphorylation of glucose is the rate-controlling step in the metabolism of glucose and that glucokinase is the valve of a pipe that funnels the flow of glucose into ATP generation via glycolysis. *GK*, glucokinase; *DHAP*, dihydroxyacetone phosphate; *GAP*, D-glyceraldehyde-3-phosphate; *NADH*, nicotinamide adenine dinucleotide, reduced

preclude a role in diabetes of a β-cell enzyme that hydrolyses G-6-P, as convincingly shown by Khan and his colleagues [27]. G-6-P, located at a four-way interchange of metabolism, has very different fates in liver and β-cells. Most striking is the virtual absence of the glycogen apparatus and of glucose-6-phosphatase in normal β-cells, and the very slow rate of metabolism via the pentose-P shunt. Glycolysis is by far the predominant pathway of G-6-P metabolism. This biochemical design feature of funnelling G-6-P almost exclusively into the glycolytic pathway is essential for β-cell glucose metabolism and glucose sensing. Isomerization to fructose-6-P is rapid and not controlled, whereas production of fructose-1,6-P_2 by P-fructokinase is probably as precisely regulated by metabolites and cofactors as in all other tissues, including liver. The regulatory role of fructose-2,6-P_2, generated by a bifunctional enzyme that also hydrolyses the metabolite, and of citrate for P-fructokinase are worth discussing here. In liver, high fructose-2,6-P_2 of about $10 \mu M$ serves as a hormone-dependent switch for turning on glycolysis and turning off gluconeogenesis. Levels of fructose-2,6-P_2 in islets are, however, an order of magnitude lower than in liver, e.g., about $1 \mu M$, and rise not more than 1.3-fold during extremes of glucose exposure, which limits its regulatory importance [2]. Citrate levels in islets are, however, substantial and in a range which could be physiologically significant. Citrate may also serve as a precursor of cytosolic acetyl-CoA which is required to explain the rise of malonyl-CoA caused by high glucose [28]. In this context it is fitting to point out that β-cell cytosol lacks fatty acid synthase [29], even though the precursors of this pathway, acetyl-CoA and malonyl-CoA, accumulate to measurable levels, again strikingly different from liver.

It is likely that fructose-6-P and fructose-1,6-P$_2$ are the most important activators of P-fructokinase and determine the glycolytic rate. The steps between fructose-1,6-P$_2$ and P-enolpruvate are near equilibrium during steady state, as is true for all tissues. Pyruvate kinase, which has a very high capacity in islet tissue, is potentially activated by fructose-1,6-P$_2$. Islet tissue lacks phosphoenolpyruvate-carboxykinase precluding reversal of glycolysis contrasting to liver cells [30]. Of great importance is the recent finding that pure populations of β-cells have negligible activity of lactate dehydrogenase [31] and the related, well established fact that islet tissue has extraordinarily high activities of mitochondrial α-glycero-P oxidase [30,31]. This design feature – high capacity of α-glycero-P oxidase and low capacity of lactate dehydrogenase – suggests that NADH generated in glycolysis is nearly quantitatively channelled to the mitochondria, rather than being used to reduce pyruvate to form lactate. The α-glycero-P shuttle is probably aided in hydrogen transport by the malate – aspartate shuttle. It is noteworthy that β-HC9 cells – a cell line that closely resembles normal β-cells – have very low cytosolic α-glycero-P dehydrogenase, which may preclude that equilibrium is established in the α-glycero-P/dihydroxyacetone-P couple (Matschinsky FM, unpublished 1995). This disproportionate endowment with the two enzymes may result in high dihydroxyacetone-P levels and may assure high substrate pressure at the glyceraldehyde-P-dehydrogenase step. Pyruvate is removed by six competing processes: efflux from the cell, reduction to lactate, reductive carboxylation to malate, transamination, oxidative decarboxylation and carboxylation. The relative importance of these six possibilities of pyruvate metabolism remains to be fully explored. It should be apparent from this brief account of the cytosolic reactions of glucose metabolism that the β-cell is geared to generate ATP, NADH and pyruvate at a rate directly proportional to glucose and almost entirely determined by the activity and glucose K_m of glucokinase, provided cytosolic ATP is plentiful.

Our knowledge of mitochondrial metabolism in β-cells is limited because the amount of mitochondria that can be isolated from islets is too small to allow the detailed study of β-cell mitochondria with classical methods. There are some obvious differences to liver, most importantly the likely absence of component steps of the urea cycle, of gluconeogenesis and of ketogenesis. However, β-cell mitochondria catalyse the important anaplerotic reaction of pyruvate carboxylation and have a high capacity of α-glycero-P oxidase (as discussed above). The capacity for fatty acid oxidation is significant. It is widely believed that Ca^{2+} regulates the activity of pyruvate-, isocitrate- and α-ketoglutarate dehydrogenase reactions of the citric acid cycle. Experimental evidence in favour of this view has been presented [32]. It is, however, reiterated in this context that the stimulation of respiration by glucose does not require extracellular Ca^{2+}, which speaks against a primary role of Ca^{2+} as a signal for activation of electron transport in β-cells. The β-cell mitochondria also seem to generate sufficient citrate as a precursor for malonyl-CoA synthesis, which is stimulated by high glucose. It is not known whether this glucose effect is Ca^{2+} dependent. Enhanced citrate synthesis due to high glucose could be independent of pyruvate dehydrogenase and might be driven by pyruvate carboxylase and acetyl-CoA that stems from noncarbohydrate sources. There is evidence that oxidative phosphorylation of β-cells is indistinguishable from that in mitochondria from many other tissues. In mitochondria, isolated from β-HC9 or β-TC3 cells, respiration and P/O ratios with glutamate, malate/pyruvate, succinate or α-glycero-P under saturation conditions were comparable to that of other mitochondria (F.M. Matschinsky and N. Doliba, unpublished).

Glucokinase Determines Metabolic Flux and Glycolytic as well as Mitochondrial ATP Generation Through Mass Action

The information in the preceding paragraphs provides the basis for the thesis that glucose governs the rate of glycolysis and respiration and that glucokinase is the flux determinant for this complex system. In contrast to acceptor control of ATP production in mitochondria by ADP, as it predominates in most tissues, glucose metabolism in the β-cell is controlled by substrate levels and is ultimately determined by blood glucose via glucokinase and metabolite coupling to cytochrome oxidase. Glucokinase appears to override any feedback control at P-fructokinase or pyruvate kinase through increasing levels of fructose-6-P and fructose-1,6-P_2 and thus governs levels of cytosolic NADH and pyruvate. Cytosolic NADH is in equilibrium with mitochondrial NADH and (reduced) flavin adenine dinucleotide (FADH) through hydrogen shuttles. The pathway of oxidative phosphorylation from these two hydrogen-carrying cofactors to cytochrome C functions near equilibrium and cytochrome oxidase activity and ATP generation is governed by cytochrome C^{2+} levels [33]. Equations 1-3 characterize this relationship:

$$1/2 \text{ NADH} + \text{cytochrome } C^{3+} + \text{ADP} + P_i = 1/2 \text{ NAD} + \text{cytochrome } C^{2+} + \text{ATP} \quad (1)$$

$$K_{eq} = \left(\frac{[NAD^+]}{[NADH]} \right)^{1/2} \frac{[C^{2+}]}{[C^{3+}]} \frac{[ATP]}{[ADP][P_i]} \quad (2)$$

$$\left(\frac{[NADH]}{[NAD^+]} \right)^{1/2} = \frac{[C^{2+}]}{[C^{3+}]} \frac{[ATP]}{[ADP][P_i]} \frac{1}{K_{eq}} \quad (3)$$

The higher the NADH/NAD ratio the higher the concentration of cytochrome C^{2+}, and hence the cytochromic oxidase activity and the ATP mass action ratio increase. The mitochondrial and cytosolic mass action ratios of ATP are probably in equilibrium. These considerations are the basis for formulating a minimal model of the metabolic and secretory events in β-cells.

Coupling of Metabolism and Insulin Secretion

Cytosolic Ca^{2+} is governed by metabolic factors and by membrane events and the cytosolic Ca^{2+} levels in turn control insulin secretion [34-36]. Ca^{2+} may also influence mitochondrial dehydrogenases, thus augmenting the total response to a fuel load. The relationships are illustrated (Fig. 2) by idealized tracings of cytosolic free Ca^{2+}, NAD(P)H fluorescence and the membrane potential from intact perifused isolated islets [6,37,38]. Increasing external glucose causes reproducible multiphase electrical and intracellular Ca^{2+} responses, which results in stimulation of insulin release. In

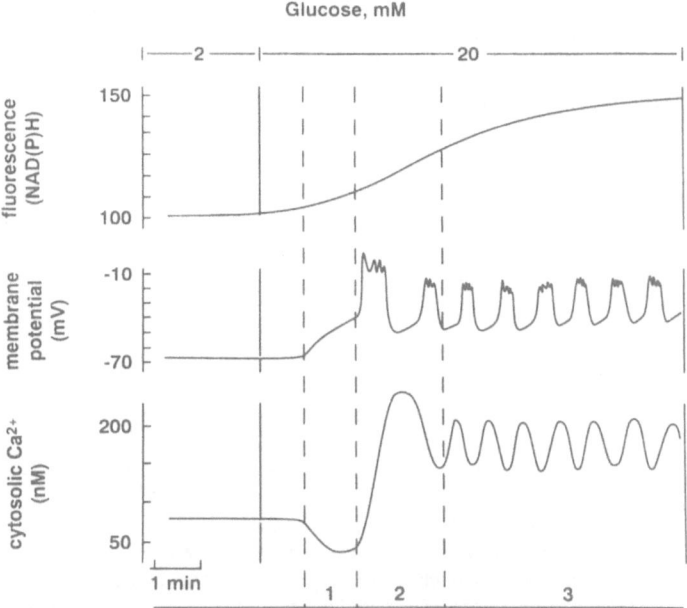

Fig. 2. *Multiphase manifestations of metabolic coupling in intact perifused islets.* The three-panel schematic figure was constructed on the basis of data presented in references [6,37,38] and of unpublished data from the author's laboratory (F.M. Matschinsky, Z. Gao, H.W. Collins and B. Wolf). For further explanation see text

phase 1, lasting for several minutes, Ca^{2+} is lowered by 25%–50%, as first observed in single β-cells by Hellman and his colleagues [39] and confirmed by many laboratories, including that of the author, with intact perifused islets. During phase 1 the membrane potential shows a slow rate of depolarization. This is followed by rapid full depolarization and a 2–5-fold rise of Ca^{2+} (phase 2, lasting 2–4 min) which passes into continuous oscillatory patterns of the electrical and Ca^{2+} tracings (phase 3). Nicotinamide adenine dinucleotide phosphate, reduced (NAD(P)H) fluorescence increases monotonically during the transition from basal levels of Ca^{2+} through phases 1–3. Remarkably, this NAD(P)H fluorescence or metabolic response is independent of extracellular Ca^{2+}. It is appreciated, however, that the monophasic NAD(P)H fluorescence increase does not preclude oscillatory fluorescence patterns at the single cell level as described by Pralong et al. [40,41]. Metabolic events of individual cells may not be synchronized in contrast to electrical phenomena and Ca^{2+} metabolism. When Ca^{2+} reaches high levels (phases 2 and 3) pyruvate metabolism may be activated to contribute to NAD(P)H generation and ATP production. This would be consistent with the observation that pyruvate by itself is not a good secretagogue but that pyruvate's efficacy to stimulate insulin release is enhanced by Ca^{2+} mobilizers, as for instance by carbachol (W.F. Pralong and C.B. Wollheim, personal communication).

   The data lead to the formulation of a hypothesis of metabolic coupling in glucose stimulated insulin release: the glucose level and glucokinase activity determine the rate of ATP, NADH and pyruvate generation in the cytosol. Hydrogen is shuttled to

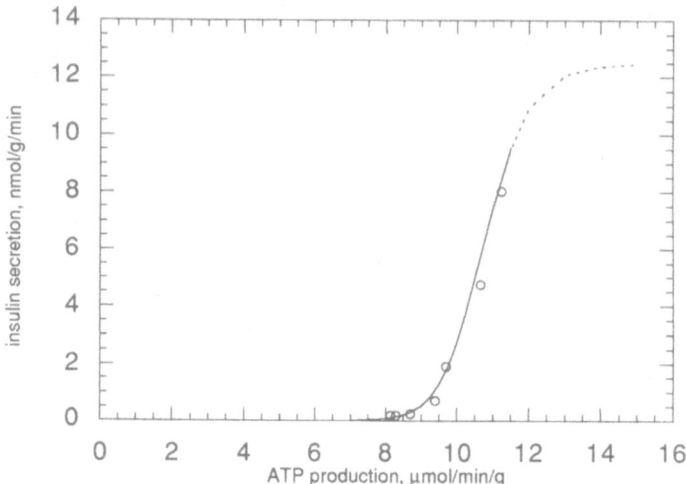

Fig. 3. *Insulin secretion from isolated rat islets as a function of the rate of ATP production.* ATP production was calculated as the unstimulated rate of ATP production (estimated as $8\,\mu$mol/min/g) plus seven times the rate of glucose usage measured at the indicated glucose concentration by release of ^3H_2O from [5-^3H]glucose. The *solid line* represents the plot of the model of Sweet and Matschinsky (in preparation). The *dashed line* represents the extrapolations to saturation with glucose which cannot be ascertained from the data

the mitochondria, leading to the generation of 6–8 mol of total ATP per mol of glucose depending on the efficacy of different shuttles (Fig. 3). Glucose thus increases ATP production about 50% above total rates. The concurrent increase of the ATP mass action ratio might be sufficient to activate Ca^{2+} ATPases and cause lowering of cytosolic Ca^{2+} (phase 1 of the Ca^{2+} tracing) and to initiate depolarization of the β-cell by inhibiting K^+ channels (phase 1 of the electrical tracing). When a threshold potential has been reached, Ca^{2+} channels open, as indicated by phase 2 of the Ca^{2+} tracing and the electrical record. The Ca^{2+} surge may activate mitochondrial dehydrogenases such that pyruvate can be completely oxidized by the citric acid cycle, which may curb the metabolism of endogenous fatty acids and amino acids. The cessation of fatty acid oxidation might be aided by the rise of malonyl-CoA, a powerful inhibitor of acyl-CoA-carnitine transferase. The production of NADH and of ATP may be further augmented. This is dramatically shown by the increase of O_2 consumption with high glucose. It should be realized, however, that the stimulation of oxygen consumption by glucose can probably be explained by the oxidation of cytosolic NADH and it is likely that pyruvate merely replaces the endogenous substrates that serve as fuel for basal respiration. Such an interpretation is strengthened by the differential effect of omitting extracellular Ca^{2+} on glucose stimulation of respiration and glucose oxidation. The activation of ion fluxes, exocytosis, anabolic and biosynthetic reactions increases the demand for ATP and augments its turnover such that the β-cell operates at a higher metabolic rate when exposed to high glucose [4]. The biological response may be further augmented by auxiliary amplifying mechanisms; for example, elevation of cAMP due to Ca^{2+} [35], activation of ASCI (ATP sensitive-calcium independent)-PLA$_2$ causing generation of arachidonic acid [42] and activation of the purinergic receptor pathway by ATP co-secreted with insulin [4].

Glucokinase Mutants Explain the Diabetic Phenotype in MODY Patients

Because it was recognized that glucokinase played an essential role as a glucose sensor as outlined in the preceding paragraphs it was predicted that even minor changes of the enzyme activity could explain certain forms of diabetes mellitus in man [1–3]. For example, a reduction of the enzyme by about 25% would cause a drastic rightward shift of the glucose dose – response curve for insulin release and would change the threshold level of the β-cells for glucose such that basal glucose might rise from normally 4–5 mM to 7–10 mM in the affected cases of diabetes. This prediction was proven to be true by the findings of human geneticists.

Froguel and his colleagues found linkage of the diabetic phenotype to glucokinase in families which show the MODY syndrome in several generations [21]. The disease

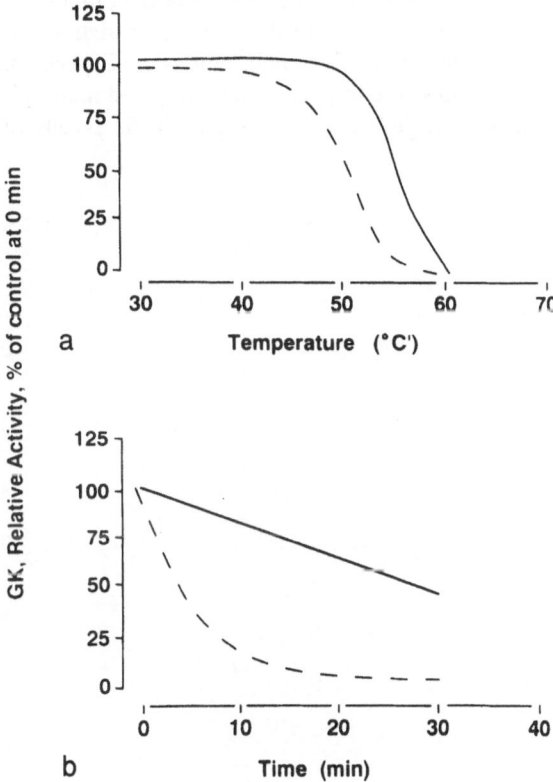

Fig. 4a,b. *Alteration of heat stability of glucokinase by the glu-300-lysine mutation of maturity-onset diabetes of the young (MODY).* Recombinant human wild-type (*solid line*) and mutant (*dashed line*; glu-300-lysine) β-cell glucokinase was incubated at different temperatures for 30 min or at 55°C for different lengths of time and the activity measured by a spectrophotometric assay. The figure is an idealized rendering of published data from our laboratory [43]

is characterized by mild hyperglycaemia with very early onset. The patients are not obese. All patients are heterozygotic for the marker. Homozygocity is most likely lethal. The defect manifests itself in impaired glucose stimulation of insulin release and probably also interferes with glucose metabolism of the liver because of the important role of glucokinase in intermediary metabolism in these two tissues. Froguel, in collaboration with the laboratory of G. Bell, succeeded in identifying many of the genetic mutations that cause the defects [21]. S. Pilkis et al. and our laboratory have investigated the effect of these mutations on glucokinase activity using recombinant human liver and β-cell isoenzyme [22,23,43]. In most instances the enzyme activity was greatly reduced due to lowered V_{max} and/or increased K_ms for glucose or ATP. The Hill number was decreased in some instances and enzyme stability appeared to be lowered in others (Fig. 4). Altogether this provides solid proof for the proposal that the glucokinase mutations are responsible for a large percentage of patients with the MODY phenotype and further strengthens the glucokinase glucose sensor concept.

The known mutation in humans affects the structure and function of the two hepatic and of the single β-cell glucokinase isoforms alike. It is therefore difficult to evaluate the relative contribution of the different organs that are likely afflicted, i.e., the role of the liver and the pancreatic islets in the MODY syndrome. Attempts are therefore being made to alter glucokinase specifically in the pancreatic β-cell or liver and to assess the impact of these targeted manipulations. Shimon Efrat introduced a new hybrid gene into mouse pancreatic β-cells to code for glucokinase antisense-

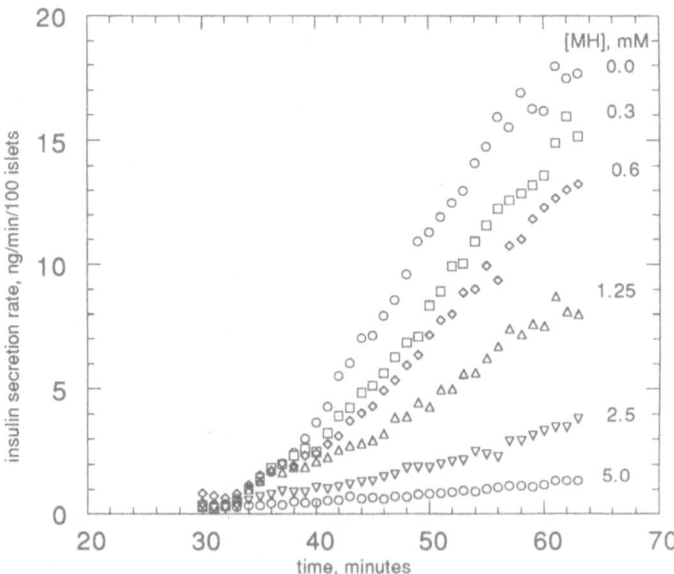

Fig. 5. *Control strength of glucokinase for glucose induced insulin release.* The effect of mannoheptulose on insulin secretion in response to a glucose ramp was studied with isolated perifused rat islets. Glucose was varied from 0 glucose at 30 min to 23 mM at 60 min in a linear fashion. The choice of the mannoheptulose levels as indicated was guided by the known K_i of about 0.7 mM

ribozymal RNA [44]. Glucokinase in islets from these transgenic mice was reduced by about 70%. However, the mice remained normoglycaemic. Nevertheless, the glucose dose – response curve of insulin release studied with the isolated perfused pancreas was shifted to the right. The effect was, however, less dramatic than expected from the measured reduction of glucokinase. These data suggest that diabetes in the human MODY cases might be due to the combined deficiency of liver and islet glucokinase.

Contrasting with the observations in transgenic mice which raise important questions about the role of glucokinase in glucose homeostasis, our recent detailed studies with mannoheptulose are designed to again document that even minute reductions of glucokinase decrease glucose responsiveness of β-cells (Fig. 5). Using the known mannoheptulose K_i of 0.7 mM and assuming equalization of intra- and extracellular heptose during a preincubation phase, it is predicted that 0.35 mM mannoheptulose should reduce glucose phosphorylation and glycolysis by about 25%. This was verified by measuring glycolysis and glucose oxidation. This also assumes that mannoheptulose has no other actions in addition to blocking glucokinase. Inhibition of glucose-stimulated insulin release was indeed demonstrable at this low degree of glucokinase inhibition as the hypothesis postulates. Much work will be needed to fully explore the MODY patients and the new transgenic animal models related to MODY that are being developed by Efrat and by other laboratories.

Outlook

Studies of the biochemical basis of glucose-induced insulin release from β-cells resulted in the formulation of the glucokinase glucose sensor hypothesis. The hypothesis explains many of the characteristics of this process, most importantly specificity and concentration dependency of hexose stimulation. It also explains the development of diabetes in a large subgroup of MODY patients. The hypothesis makes many important postulates which await further testing:

1. Is there indeed direct transmission of the secretory signal from glucose to cytochrome oxidase by mass action?
2. Are there impairments of glycolysis by the glucokinase regulatory protein or glucose-6-phosphatase in diabetes mellitus?
3. What is the biochemical basis of glucokinase induction by increased extracellular glucose?
4. Is there adaptation of glucokinase activity of β-cells in type II diabetes mellitus and, if so, to what extent and to what advantage or disadvantage?
5. What is the proposed role of glucokinase in extrahepatic and extrapancreatic cells?
6. What is the physiological role of the powerful in vitro inhibition of glucokinase activity by long chain acyl-CoA?
7. What are the detailed physico-chemical processes of glucokinase catalysis of glucose phosphorylation and of the inhibition of this process by the regulatory protein and acyl-CoA?
8. What is the significance, if any, of the newly discovered linkage of IDDM to the glucokinase gene?

9. Are there ways to activate glucokinase isoenzymes by pharmacological intervention and, if so, is such activation beneficial in diabetes mellitus?

This is a small selection of questions that emanate from the glucokinase glucose sensor hypothesis which will undoubtedly engage many investigators in the years to come.

References

1. Meglasson MD, Matschinsky FM (1984) New perspectives on pancreatic islet glucokinase. Am J Physiol 246 (Endocrinol Metabol 9): E1–E13
2. Meglasson MD, Matschinsky FM (1986) Pancreatic islet glucose metabolism and regulation of insulin secretion. DiabetesMetab Rev 2: 163–214
3. Matschinsky FM (1990) Glucokinase as glucose sensor and metabolic signal generator in pancreatic β-cells and hepatocytes. Diabetologia 39: 647–652
4. Erecinska M, Bryla J, Michalik M, Meglasson MD, Nelson D (1992) Energy metabolism in islets of Langerhans. Biochim Biophys Acta 1101: 273–295
5. Hutton JL, Malaisse WJ (1980) Dynamics of O_2 consumption in rat pancreatic islets. Diabetologia 18: 395–405
6. Gilon P, Henquin JC (1992) Influence of membrane potential changes on cytoplasmic Ca^{2+} concentration in an electrically excitable cell, the insulin secreting pancreatic β-cell. J Biol Chem 267: 20713–20720
7. Matschinsky FM, Ellerman JE (1968) Metabolism of glucose in islets of Langerhans. J Biol Chem 243: 2730–2736
8. Grodsky GM, Batts AH, Bennett LL, Vcella C, McWilliams NB, Smith DF (1963) Effects of carbohydrates on secretion of insulin from isolated rat pancreas. Am J Physiol 205: 638–644
9. Coore HG, Pandle PJ (1964) Regulation of insulin secretion studied with pieces of rabbit pancreas incubated in vitro. Biochem J 93: 66–78
10. Malaisse WJ, Lea MA, Malaisse-Lagae F (1968) The effect of mannoheptulose on the phosphorylation of glucose and the secretion of insulin by islets of Langerhans. Metabolism 17: 126–132
11. Ashcroft SJH, Randle PJ (1970) Enzymes of mouse pancreatic islets. Biochem J 119: 5–15
12. Renold AE (1970) Insulin biosynthesis and secretion – a still unsettled topic. N Engl J Med 282: 173–182
13. Ashcroft SJH, Bassett JM, Randle PJ (1972) Insulin secretion mechanisms and glucose metabolism in isolated islets. Diabetes 21 [Suppl 2]: 538–545
14. Matschinsky FM, Ellerman JE, Stilling S, Raybaud F, Pace C, Zawalich WS (1975) Hexoses and insulin secretion. In: Hasselblatt A, Bruchhausen FV (eds) Insulin II. Springer Verlag, Berlin Heidelberg New York, pp 79–114 (Handbook of experimental pharmacology, vol 32)
15. Malaisse WK, Sener A (1985) Glucokinase is not the pancreatic beta-cell glucoreceptor. Diabetologia 28: 520–527
16. Magnuson MA (1990) Glucokinase gene structure. Functional implications of molecular genetic studies. Diabetes 39: 523–527
17. Iynedjian PB (1993) Mammalian glucokinase and its gene. Biochem J 293: 1–13
18. Newgard CB, Ferber S, Quaade C, Johnson JH, Hughes SD (1994) Molecular engineering of glucose regulated insulin secretion. In: Drasnin B, LeRoith D (eds) Molecular biology of diabetes, vol. I. Humana, Totowa, pp 119–154
19. Permutt MA, Chiu KC, Tanizawa Y (1992) Glucokinase and NIDDM: a candidate gene paid off. Diabetes 41: 1367–1372
20. Tanizawa Y, Matsutani A, Chiu KC, Permutt MA (1992) Human glucokinase gene: isolation, structural characterization and identification of a microsatellite repeat polymorphism. Mol Endocrinol 6: 1070–1081
21. Froguel P, Zouali H, Vionnet N, Velho G, Vaxillaire M, Sun F, Lesage S, Stoffel M, Takeda J, Pasa P, Permutt A, Beckmann J, Bell G, Cohen D (1993) Familial hyperglycemia due to mutations in glucokinase. N Engl J Med 328: 697–702

22. Pilkis SJ, Weber IT, Harrison RW, Bell GI (1994) Glucokinase: structural analysis of a protein involved in susceptibility of diabetes. J Biol Chem 269: 21925–21928

23. Matschinsky FM, Liang Y, Kesavan P, Wang L, Froguel P, Velho G, Cohen D, Permutt MA, Tanizawa Y, Jetton TL, Niswender K, Magnuson MA (1993) Glucokinase as pancreatic β-cell glucose sensor and diabetes gene. J Clin Invest 92: 2092–2098

24. Rowe RE, Walpaehorst B, Bell GI, Rish N, Spielman RS, Concannon P (1995) Linkage and association between insulin dependent diabetes mellitus (IDDM) susceptibility and markers near the glucokinase gene on chromosome 7. Nat Genet 10: 240–242

25. Heimberg H, DeVos A, Vandercammen A, von Schaftingen E, Pipeleers D, Schuit F (1993) Heterogeneity in glucose sensitivity among pancreatic β-cells is correlated to differences in glucose phosphorylation rather than glucose transport. EMBO J 12: 2873–2879

26. Perales MA, Sener A, Malaisse WJ (1991) Hexose metabolism in pancreatic islets: the glucose-6-phosphatase riddle. Mol Cell Biochem 101: 67–71

27. Khan AV, Chaudramouli V, Gotenson CG, Ahren B, Schumann WC, Low H, Landau BR, Efendic S (1989) Evidence for presence of glucose cycling in pancreatic islets of the ob/ob mouse. J Biol Chem 264: 9732–9733

28. Liang Y, Matschinsky FM (1991) Content of CoA-esters in perifused rat islets stimulated by glucose and other fuels. Diabetes 40: 327–333

29. Brun T, Assimacopoulos-Jeannet F, Roche E, Prentki M (1992) Malonyl-CoA in β-cell signalling: control of fatty acid oxidation or substrate for lipid biosynthesis? Diabetologia 35 [Suppl 1]: A13–40

30. MacDonald MJ (1990) Elusive proximal signals of β-cells for insulin secretion. Diabetes 39: 1461–1366

31. Sekine N, Cirulli V, Regazzi R, Brown LJ, Gine E, Tamarit-Rodriguez J, Girotti M, Marie S, MacDonald MJ, Wollheim CB (1994) Low lactate dehydrogenase and high mitochondrial glycerol phosphate dehydrogenase in pancreatic β-cells. J Biol Chem 269: 4895–4902

32. McCormack JG, Lango EA, Corkey BE (1990) Glucose induced activation of pyruvate dehydrogenase in isolated rat pancreatic islets. Biochem J 267: 527–530

33. Voet D, Voet JG (1990) Electron transporter and oxidative phosphorylation. In: Voet D, Voet JG (eds) Biochemistry. Wiley, Chichegter, pp 528–557

34. Wollheim CB, Sharp GWG (1981) Regulation of insulin release by calcium. Physiol Rev 61: 914–973

35. Prentki M, Matschinsky FM (1987) Ca^{2+}, cAMP and phospholipid derived messengers in coupling mechanisms of insulin secretion. Physiol Rev 67: 1185–1248

36. Ashcroft FM, Ascroft SJH (1992) Mechanisms of insulin secretion. In: Ashcroft FM, Ascroft SJH (eds) Insulin. Oxford University Press, Oxford, pp 97–150

37. Roe MW, Lancaster ME, Mertz RJ, Worley JF III, Dukes ID (1993) Voltage-dependent intracellular calcium release from mouse islets stimulated by glucose. J Biol Chem 268: 9953–9956

38. Dukes ID, McIntyre MS, Mertz RJ, Philipson LH, Roe MW, Spencer B, Worley JF III (1994) Dependence on NADH produced during glycolysis for β-cell glucose signalling. J Biol Chem 269: 10979–10982

39. Hellmann B, Gylfe E, Grappengiesser E, Lund PE, Marcström A (1992) Cytoplasmic calcium and insulin secretion. In: Flatt PR (ed) Nutrient regulation of insulin secretion. Portland, London, pp 213–246

40. Pralong WF, Bartley C, Wollheim CB (1990) Single islet β-cell stimulation by nutrients: relationship between pyridine nucleotides, cytosolic Ca^{2+} and secretion. EMBO J 9: 53–60

41. Pralong WF, Spat A, Wollheim CB (1994) Dynamic pacing of cell metabolism by intracellulor Ca^{2+} transient. J Biol Chem 269: 27310–27314

42. Turk J, Gross RW, Ramanadham S (1993) Amplification of insulin secretion by lipid messengers. Diabetes 42: 367–374

43. Liang Y, Kesavan P, Wang L, Niswender K, Tanizawa Y, Permutt MA, Magnuson MA, Matschinsky FM (1995) Variable effects of MODY-associated glucokinase mutations on substrate interactions and stability of the enzyme. Biochem J 309: 167–173

44. Efrat S, Leiser M, Wu Y-J (1994) Ribozyme-mediated attenuation of pancreatic B-cell glucokinase expression in transgenic mice results in impaired glucose-induced insulin secretion. Proc Natl Acad Sci USA 91: 2051–2055

How Ca^{2+} and Other Signalling Pathways Control the Exocytosis of Insulin in the β-Cell

C.B. WOLLHEIM

Prologue

Albert E. Renold founded the Institut de Biochimie Clinique of which he was the director from 1963 to his untimely death in 1988. He created a unique environment for experimental diabetes research that attracted junior and senior scientists from all over the world. Until its integration into the modern University Medical School building in 1985, the Institut was housed in a villa from the last century. The crowded, quite primitive laboratories of the old Institut surprised every visitor. The inspiring leadership and continuous encouragement of our director helped us to improvise and extend our activities by squatting other laboratories more suitable for certain experiments. In this way we profited from the tissue culture facilities of the Department of Morphology with whose director, Lelio Orci, Albert Renold had close personal and professional ties.

Albert Renold exerted a great influence on diabetes research. This immediately became apparent to me when in 1971 I joined his group as a junior research fellow. I had the privilege to participate in a symposium on insulin biosynthesis and secretion organized by him in Fribourg (Switzerland) only 3 weeks after my arrival. Virtually all leading scientists of experimental and clinical diabetology had accepted the invitation. Albert Renold's impact on the field and his popularity were also reflected by the many invited professors to the Institut, who endured the hardship of sharing office and laboratory space with an ever-growing crowd of international scientists. Albert Renold strongly believed that the mixing of trainee researchers with scholars in a confined environment is most fruitful for both categories. Indeed, we all remember the extraordinary atmosphere of intense discussions of every new experimental result of the group or in the literature. Albert Renold was much more than an enthusiastic and impressive research director. He engaged himself with unusual compassion in our personal preoccupations which made us all feel part of his own devoted family.

Albert Renold's many contributions to the field are highlighted by the abbreviated list of his publications printed in this book. During his 25 years in Geneva, his three main research interests were the control of insulin biosynthesis and secretion, the elucidation of the action of insulin, and the study of animal models of diabetes mellitus. His investigative mind early defined the relevant questions on how insulin secretion is regulated [1]. The problem of the coupling of glucose metabolism to insulin secretion and the implication of Ca^{2+}, cAMP and the mode of action of the inhibitory neurohormone epinephrine were intensively studied by him [1]. The aim of this article is to illustrate the recent development of our knowledge in the field of insulin secretion. I am sure that Albert Renold would be pleased and perhaps amazed by the results obtained in the area of insulin secretion through the application of the novel techniques in cellular and molecular biology to his beloved β-cell.

Introduction

The moment-to-moment adaptation of insulin secretion from the pancreatic β-cell to the fluctuations of glucose and other nutrients in the blood circulation is the single most important factor controlling blood glucose homeostasis. Incapacity of the β-cells to increase rapidly the rate of insulin release in anticipation of food intake and during its absorption characterizes many forms of type 2 or non-insulin-dependent diabetes mellitus [2,3]. Stimulus-secretion coupling of insulin release evoked by glucose is characterized by several steps. These include: (1) stimulus recognition by the metabolic degradation of glucose, following its uptake by facilitated diffusion; (2) generation of metabolic coupling factors mainly by the mitochondria; (3) changes in ion fluxes at the plasma membrane leading to membrane depolarization and Ca^{2+} influx; (4) transport of the insulin-containing secretory granules to the plasma membrane and the fusion of the granule and plasma membranes. In view of the complex molecular basis of each of these four steps, it is not surprising that impaired insulin secretion is a common feature in diabetes mellitus. Although considerable effort has been spent on the investigation of the genetics of type 2 diabetes, so far only rare forms of the disease have been linked to mutations in key proteins implicated in one of the four steps in stimulus-secretion coupling. Thus, defects in the glucose-phosphorylating enzyme, glucokinase, the rate-limiting enzyme reaction for glucose metabolism in the β-cell, have been causally linked to one of the subforms of maturity onset diabetes of the young (MODY) [4]. Mitochondrial diabetes, characterized by point mutations or deletions in the maternally inherited mitochondrial DNA, has also been associated with impaired insulin secretion [5].

Insulin is stored in secretory granules corresponding to the large dense core secretory vesicles of neurons and neuroendocrine cells that contain peptide neurotransmitters and hormones. Both types of vesicle are assembled at the *trans* side of the Golgi complex [6]. The β-cell contains approximately 13 000 granules, of which only a small fraction is released even during strong stimulation [7]. The transport of the granules to the plasma membrane seems to involve microtubule-associated motor proteins and interruption of microtubules attenuates mainly the second or sustained phase of insulin secretion [8]. Actin-containing microfilaments have been proposed both to constitute a barrier blocking the access of the granules to the plasma membrane [8] and to be required for normal recruitment of the granules to the cell boundary [9].

As other endocrine cells, the β-cell belongs to the "classical" secretory cell type [10]. These cells display regulated secretion, characterized by the tight control of exocytosis. This is the final step of the secretory pathway [6], comprising docking of the granules to the plasma membrane, the fusion of the granule and plasma membranes and the fission of the granules [6,10,11]. The other type of secretion, constitutive secretion, occurs in all cells [10]. It is controlled at the step of vesicle budding from the *trans*-Golgi network [12]. Ca^{2+} and cAMP, the main signals that trigger regulated secretion, are not primarily implicated in constitutive secretion. However, this does not imply distinct mechanisms underlying the two forms of secretion. On the contrary, it is now well established that exocytosis is regulated in a manner similar to intracellular membrane fusion events and that this process has been conserved from yeast to mammals [13,14]. Our understanding of membrane fusion has undergone considerable progress in recent years through the knowledge gained from cell-free membrane fusion assays and from yeast genetics. The aim of this article is to summarize available information on the exocytosis of insulin and how the model proposed for neurotransmitter secretion can be applied to the β-cell [15].

Direct Control of the Exocytosis of Insulin by Intracellular Signalling Pathways

Generation of the Ca^{2+} Signal by Glucose

It has been known for more than three decades that glucose and other nutrient stimuli only elicit insulin secretion in the presence of physiological extracellular Ca^{2+} concentrations both in vivo and in vitro (for a review see [7]). This Ca^{2+} dependency explains the impairment of insulin secretion in states of hypocalcaemia, such as vitamin D deficiency [16]. These and other findings favoured a second messenger role for Ca^{2+} in stimulus-secretion coupling in the β-cell. There are two reasons why Ca^{2+} is a universal intracellular messenger molecule. First, its concentration in the cytosol ($[Ca^{2+}]_i$) can rapidly be increased since the resting $[Ca^{2+}]_i$ is kept around $100\,nM$, i.e., there is an inwardly directed electrochemical gradient for Ca^{2+} of $10\,000$ [17]. Second, all cells express Ca^{2+} receptor or binding proteins with Ca^{2+} affinities appropriately adapted to the $[Ca^{2+}]$ of the given cellular compartment [17]. In the β-cell and other excitable cells, $[Ca^{2+}]_i$ is increased by the opening of voltage-sensitive Ca^{2+} channels which promotes Ca^{2+} influx [18]. Another means of increasing the passive influx of Ca^{2+} into the cytosol is the opening of Ca^{2+} release channels in the endoplasmic reticulum, a Ca^{2+} store which has an equally high $[Ca^{2+}]$ as the extracellular fluid [17,19]. Ca^{2+} mobilization is mediated by inositol(1,4,5)trisphosphate (InsP$_3$) binding to its receptor, a Ca^{2+} channel in the membrane of the endoplasmic reticulum [17,20]. Acetylcholine and cholecystokinin activate phospholipase C and promote Ca^{2+} mobilization by this mechanism [21], while glucose only generates InsP$_3$ as a consequence of Ca^{2+} influx [22].

Although it was shown already in the late 1960s that glucose depolarizes the β-cell membrane and causes electrical activity which could elicit a rise in $[Ca^{2+}]_i$ [7], the underlying mechanism was only discovered in 1984 with the demonstration that glucose closes ATP-sensitive K^+ channels [23]. These channels are also closed by hypoglycaemic sulphonylureas such as tolbutamide. Whereas glucose and leucine, another physiological nutrient secretagogue, raise the ATP/ADP ratio following stimulation of oxidative metabolism (see Matschinsky, this volume), the sulphonylureas bind directly to the channel. Both the ATP-sensitive K^+ channel and the associated sulphonylurea receptor have recently been cloned [24,25].

The membrane depolarization caused by nutrients and by sulphonylureas promotes Ca^{2+} influx by the gating of voltage-sensitive Ca^{2+} channels, predominantly of the L-type [18]. The measurement of the fluorescence of endogenous reduced pyridine nucleotides [NAD(P)H] is a convenient means of monitoring oxidative metabolism, as it reflects mainly changes in mitochondrial pyridine nucleotides [26,27]. The firing of action potentials (rapid, transient depolarizations) coincides with the increase in $[Ca^{2+}]_i$ and the commencement of insulin secretion. In a minority of isolated rat β-cells, glucose-elicited metabolic oscillations can be detected which appear to be driven by the $[Ca^{2+}]_i$ transients and display frequencies (ca. 2/min) [27] similar to those of the bursts of the electrical activity recorded in intact mouse islets [28]. Examples of oscillations in $[Ca^{2+}]_i$ and NAD(P)H in single, glucose-stimulated rat β-cells are shown in Fig. 1. Note that inhibition of Ca^{2+} influx by removal of extracellular Ca^{2+} with the chelator EGTA abolishes the effect of glucose on $[Ca^{2+}]_i$ and attenuates the increase in NAD(P)H fluorescence (Fig. 1C,D). The latter finding suggests that $[Ca^{2+}]_i$ increases may augment mitochondrial metabolism to ensure that ATP production balances the enhanced ATP consumption during the stimulation of exocytosis [29]. Glucose also raises the concentration of Ca^{2+} in the mitochondrial matrix during insulin secretion [30]. This in turn is thought to activate Ca^{2+}-sensitive mitochondrial dehydrogenases, promoting NADH production and increased respiration [29]. Glu-

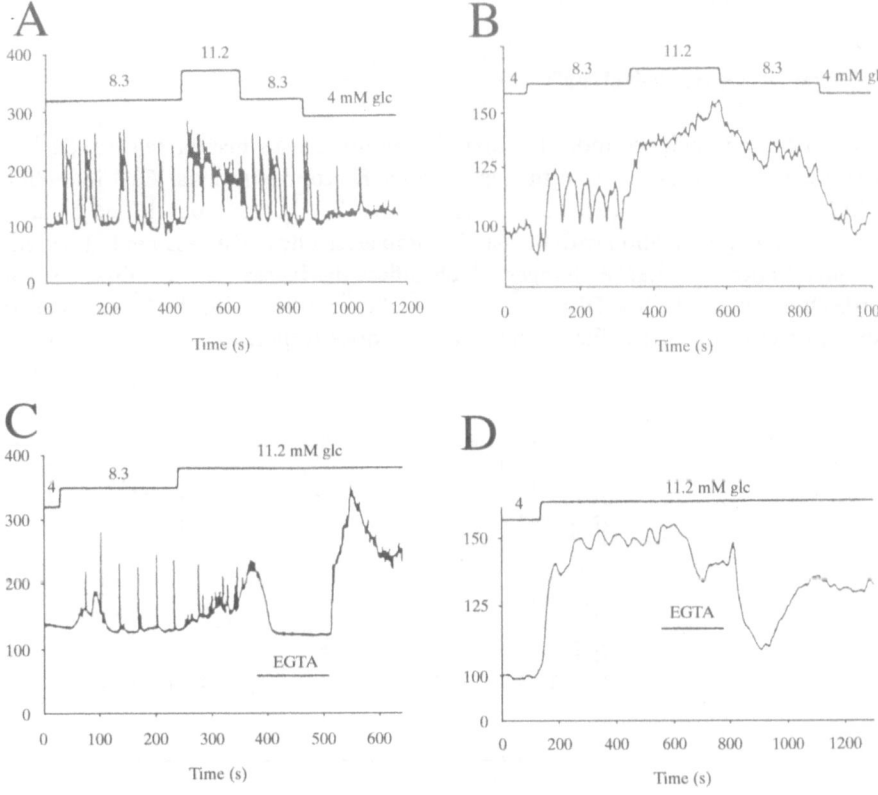

Fig. 1A–D. Effect of glucose on $[Ca^{2+}]_i$ (A,C) and NAD(P)H fluorescence (B,D) in single rat β-cells. The two parameters were recorded in parallel experiments by dual excitation wavelength microspectrofluorometry, as described elsewhere [26,27], using fura-2 for the $[Ca^{2+}]_i$ measurements. In C and D, extracellular Ca^{2+} was removed by the addition of the Ca^{2+} chelator EGTA. The basal glucose concentration of the medium was 4 mM. (Reproduced from [27] with permission)

cose and other nutrients have been found to promote insulin secretion under conditions in which $[Ca^{2+}]_i$ is clamped to a stimulatory level by K^+ depolarization. The action of the nutrients on the ATP-sensitive K^+ channel was blocked by the channel opener diazoxide. The stimulation of insulin secretion under these permissive $[Ca^{2+}]_i$ conditions is intriguing. It is most probably caused by the production of metabolic coupling factors of mitochondrial origin that reinforce the action of Ca^{2+} on exocytosis [31]. Less than 10% of isolated β-cells exhibit clear $[Ca^{2+}]_i$ oscillations of the rapid (2/min) type [27]. These cells may represent a subpopulation of pacemaker cells governing the synchronization of the activity of islet territories or entire islets. Such coupling between islet cells is discussed in Meda's contribution (this volume). In single mouse [32] and human [33] islets, the glucose-induced rapid oscillations in $[Ca^{2+}]_i$ recorded from the entire islet correlate well with spikes in insulin secretion. These findings strongly favour the idea that the rise in $[Ca^{2+}]_i$ is the main trigger for insulin secretion.

Insulin secretion measured in the effluent of the perfused pancreas of experimental animals or in peripheral blood of humans is also pulsatile, albeit with a slower frequency. The pulsatile nature of insulin secretion may be of biological importance as it is lost in patients with type 2 diabetes (for a discussion see [33]).

Ca^{2+} Stimulates the Exocytosis of Insulin

Overwhelming evidence indicates that Ca^{2+} influx causes insulin release. Such experiments were performed with depolarizing K^+ concentrations, Ca^{2+} ionophores [7,34], or tolbutamide [32]. Most elegantly this has been accomplished by direct membrane depolarization with the patch clamp electrode and assessment of secretion by membrane capacitance changes which reflect the increase in membrane surface area after exocytosis [35]. Direct stimulation of exocytosis of insulin by Ca^{2+} was first demonstrated in permeabilized islets [36]. The dose-response characteristics of Ca^{2+}-

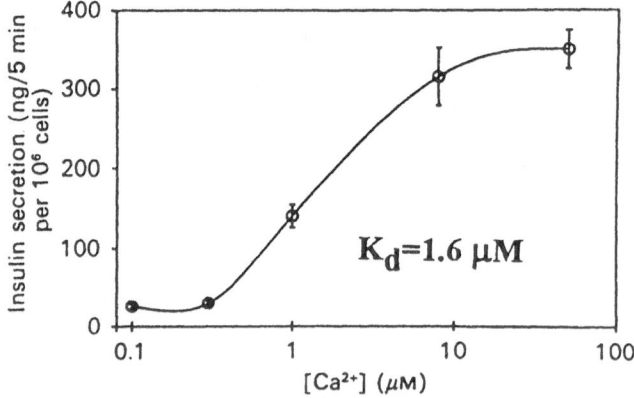

Fig. 2. Ca^{2+}-induced insulin secretion from electrically permeabilized HIT-T15 cells. Suspensions of HIT-T15 cells were permeabilized and incubated at 37°C in the presence of 5 mM ATP and the indicated Ca^{2+} concentrations, which were verified with a Ca^{2+}-selective electrode. (Reproduced from [37] with permission)

induced insulin exocytosis are shown in Fig. 2. As is the case in permeabilized islets, in the permeabilized HIT-T15 cells, a hamster β-cell line, the EC$_{50}$ is 1–2 μM Ca^{2+} [37]. This value is identical to that found in native, patch-clamped β-cells [38]. In permeabilized cells, the resting Ca^{2+} concentration is clamped to 100 nM, corresponding to the [Ca^{2+}]$_i$ of nonstimulated β-cells (see Fig. 1) [26,27,32,33]. As in intact cells, Ca^{2+}-stimulated exocytosis from permeabilized cells requires permissive temperatures (above 20°C) and is dependent on ATP. While 1 mM ATP is sufficient for optimal effects of Ca^{2+} [39], drastic depletion of ATP still allows some (10%–15%) secretion to occur [40]. It appears that ATP is required for the priming of exocytosis and that the extent of ATP-independent Ca^{2+}-stimulated exocytosis is determined by the number of secretory vesicles already docked to the plasma membrane, as established in melanotrophs and chromaffin cells [41]. Little information is available on the number of docked secretory granules in the β-cell or β-cell lines, but it appears that only a small proportion is docked in unstimulated cells [6]. Glucose seems to increase the movement of the granules towards the cell boundary, a process that has been termed margination [6]. It remains to be established whether the granules dock at specific sites corresponding to the active zones in nerve terminals. Nonetheless, colocalization of voltage-sensitive Ca^{2+} channels and cytoplasmic domains enriched in granules has been found in mouse β-cells [38]. These domains are depleted from granules on application of depolarizing membrane potential voltage jumps that promote localized Ca^{2+} influx [38].

Glucose elicits trains of Ca^{2+} action potentials which are seen as bursts of spikes during the electrical activity [23,28]. As already mentioned, insulin is secreted from single islets in pulses that are in phase with the bursts [32,33]. Whereas glucose raises [Ca^{2+}]$_i$ beneath the plasma membrane during a single action potential [42], a simulated single action potential was insufficient for the triggering of exocytosis. Rather, a train of at least four such potentials was required for the effect [35]. This suggests a [Ca^{2+}]$_i$ threshold for exocytosis and a build-up of [Ca^{2+}]$_i$ during the firing of action potentials. Gradients of [Ca^{2+}]$_i$ between the cell periphery and its interior have been demonstrated by video imaging in single, glucose-stimulated β-cells [42]. Such gradients have also been shown by measuring exocytosis in patch-clamped β-cells. Indeed, depolarization-induced localized [Ca^{2+}]$_i$ rises beneath the plasma membrane were more efficient in causing secretion than either Ca^{2+} mobilization with InsP$_3$ or Ca^{2+} infusion through the patch pipette. This study concluded that [Ca^{2+}]$_i$ at the release site reaches a concentration of several micromolar [35]. Thus, the data from permeabilized cells and single, patch-clamped cells concur extremely well. The Ca^{2+} affinity is similar to the value established for the release of large dense core secretory vesicles in other endocrine cells [41] and neurons but is two orders of magnitude lower than the 100 μM affinity of exocytosis of synaptic vesicles [11,15].

Effects of cAMP-Dependent Protein Kinase and Protein Kinase C on Insulin Exocytosis

Insulin secretion is potentiated by gastrointestinal hormones such as glucagon-like peptide-1 (GLP-1), gastric inhibitory polypeptide (GIP) and cholecystokinin (CCK). These hormones are secreted from the gut during food absorption and have been implicated in the enhanced insulin secretion evoked by oral compared to intravenous administration of glucose [43]. Acetylcholine (ACh), released from intraislet terminals of the vagus nerve already during the preabsorptive (cephalic) phase of food intake also enhances insulin secretion [7]. GLP-1, glucagon and GIP bind to receptors that couple via the heterotrimeric G-protein G_s to adenylate cyclase which promotes the generation of cAMP and activation of cAMP-dependent protein kinase (PKA) [43]. cAMP acts on both early and late events in stimulus-secretion coupling. The $[Ca^{2+}]_i$ signal is enhanced by an effect on Ca^{2+} channels (reduced inactivation) [18,44] by an increase in the proportion of β-cells that respond to glucose with K^+ channel closure and membrane depolarization, also referred to as improved glucose competence [45]. In addition, cAMP exerts an ion channel independent effect on the exocytosis process itself. Thus, Ca^{2+}-induced exocytosis is increased by cAMP both in permeabilized islets [46] and RINm5F cells [39]. Similar results have been obtained in single, patch-clamped β-cells by monitoring membrane capacitance [44]. As in intact cells [7], cAMP sensitizes insulin exocytosis to the action of Ca^{2+} and is incapable of initiating the secretory process.

ACh and CCK couple to phospholipase C, most probably via the heterotrimeric G protein G_q [47]. This leads to the production of $InsP_3$ and diacylglycerol, the latter messenger being the principal activator of protein kinase C (PKC) [48]. Stimulation of one or several of the isoforms of PKC also elicits insulin secretion both in intact and permeabilized cell preparations [48], as well as in patch-clamped single cells [49]. The molecular targets for PKA and PKC have not been identified at present [46].

The involvement in insulin secretion of another protein kinase, the Ca^{2+}/calmodulin-dependent protein kinase II (CaM kinase II) has been proposed. This enzyme is present in the β-cell and has been shown to be activated during the first phase of glucose-induced insulin secretion [50]. An inhibitory peptide of CaM kinase II was also reported to attenuate Ca^{2+}-stimulated exocytosis in single β-cells [35]. The final demonstration of the role of this kinase requires the identification of its substrates and experiments involving manipulation of the expression of the kinase in insulin-secreting cells. In islets of transgenic mice, overexpression of calmodulin, the ubiquitous Ca^{2+} receptor protein, leads to decreased insulin secretion, even when a mutant with normal Ca^{2+} binding but extremely low affinity for the target enzymes of CaM was used [51].

Role of Guanine Nucleotide Binding Proteins in Insulin Exocytosis

Pertussis Toxin-Sensitive Heterotrimeric G Proteins

Guanine nucleotide binding proteins (G proteins or GTPases) fall into two categories. The heterotrimeric G proteins are mainly, but not exclusively, localized at the inner surface of the plasma membrane and monomeric or small G proteins are expressed in many cellular compartments. The former are composed of an α-subunit which both binds and hydrolyses GTP. The $\beta\gamma$ dimer reversibly associates with the α-subunit depending on the occupancy of the hormone receptor [52,53]. In this way the hormones that stimulate adenylate cyclase couple to G_s [43,52,53] and those that activate phospholipase C to G_q [47,52,53]. The β-cell also expresses two families of pertussis toxin-sensitive heterotrimeric G proteins, G_{il-3} and G_{ol-2} [40,54].

It has been known for two decades that pertussis toxin abolishes the inhibition of insulin secretion by epinephrine and norepinephrine, an effect mediated by α_2-adrenergic receptors [7]. Pertussis toxin was therefore also called "islet activating protein" [7]. The toxin catalyses the transfer of ADP ribose from NAD to a cysteine close to the C terminus of the G_i and G_o proteins. This covalent modification makes the G proteins insensitive to the occupied hormone receptor, i.e., the hormone can no longer promote the exchange of GDP for GTP. Both pertussis toxin and cholera toxin, which ADP ribosylates G_s with its activation as a consequence, act on intact cells [52,53]. This has made them widely used tools for the probing of the involvement of the subtype of G protein in a given biological process.

Epinephrine, somatostatin and galanin, as well as other neuropeptides that inhibit insulin secretion, have been suggested to act on various steps in the stimulus-secretion coupling [37,40,55,56]. The neurohormones inhibit adenylate cyclase and lower cAMP levels and transiently decrease $[Ca^{2+}]_i$ [55,57]. The latter appears to be a consequence of both closure of L-type Ca^{2+} channels and hyperpolarization of the membrane potential following the opening of K^+ channels ([34] and references therein). However, none of these actions provides a satifactory explanation of the profound and long-lasting inhibition of stimulated insulin secretion by the neurohormones. We suggested early that epinephrine and somatostatin must act by interfering with a final step in secretion situated beyond the regulation of the second messengers cAMP and Ca^{2+} [7]. This view was supported by the findings that epinephrine inhibited cAMP-stimulated insulin secretion and that somatostatin attenuated secretion induced by the Ca^{2+} ionophore A23187 [7]. More compelling

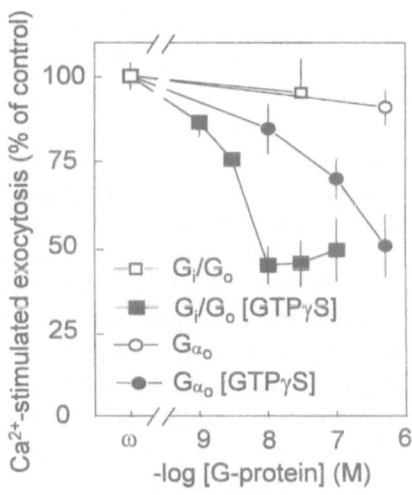

Fig. 3. Direct inhibition of Ca^{2+}-stimulated exocytosis of insulin by exogenous preactivated G_i/G_o proteins. HIT-T15 cells were permeabilized with streptolysin-O and exposed to G proteins purified from bovine brain. The G proteins were preactivated or not with 40 nM GTPγS. Secretion was stimulated five- to eightfold by raising the Ca^{2+} concentration from 0.1 μM to 10 μM. Values are expressed as percentage of Ca^{2+}-induced secretion compared to control rates. (Reproduced from [40] with permission)

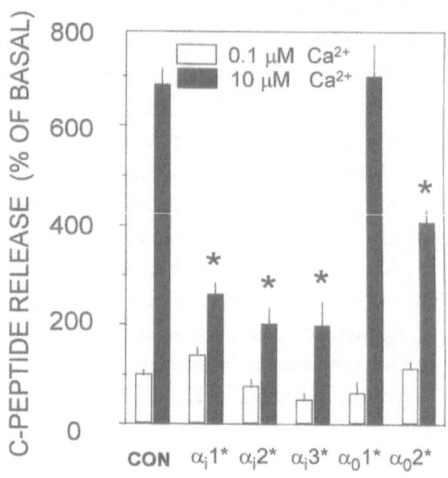

Fig. 4. Inhibition of Ca^{2+}-stimulated exocytosis of insulin by constitutively active mutants of G$_{\alpha i}$ and G$_{\alpha o}$. HIT-T15 cells were transiently transfected with cDNA-encoding human proinsulin or cotransfected with this plasmid and cDNA encoding the indicated, constitutively active mutants of G$_{\alpha i}$ and G$_{\alpha o}$ isoforms. Forty-eight hours after transfection, cells were permeabilized with streptolysin-O and subsequently exposed to low or high Ca^{2+} concentrations. Exocytosis from the transfected cells was monitored by measuring human C-peptide secretion, as detailed in [40]

evidence for a direct inhibition of exocytosis of insulin came from experiments in permeabilized cell preparations. It was first shown in permeabilized whole islets that epinephrine inhibits Ca^{2+}-stimulated insulin secretion [58]. In electrically permeabilized RINm5F cells, it could clearly be demonstrated that the inhibition of exocytosis by epinephrine is completely independent of the regulation of cAMP levels [55]. Moreover, in HIT-T15 cells permeabilized with streptolysin-O (SLO), which creates large holes in the plasma membrane and loss of cytosolic proteins, epinephrine was still able to attenuate insulin exocytosis evoked by Ca^{2+} [40]. As shown in Fig. 3, the addition of G$_i$/G$_o$, or the α-subunit of G$_o$ also attenuates insulin exocytosis [40]. The merely partial inhibition of secretion by epinephrine in permeabilized cells could be due to the use of buffers with an intracellular ion composition (high K$^+$ concentration) which is unfavourable for the binding of epinephrine to its receptor. Indeed, when receptor mimetic peptides were used to bypass the step of hormone binding and to directly activate G$_i$ or G$_o$, the inhibition was near complete [40]. Finally, transient overexpression of the α-subunit of the three G$_i$ subtypes and of G$_{o2}$ caused marked attenuation of insulin secretion both from intact and permeabilized cells (Fig. 4). The inhibition is not due to untoward effects of overexpression of G proteins, as the α-subunit of G$_{o1}$ did not affect Ca^{2+}-induced exocytosis [40]. Taken together, these data clearly demonstrate that inhibitory neurohormones exert a G-protein-mediated direct action on the exocytosis of insulin. Whether this effect is due to blockade of the putative fusion pore (see below) or to stimulation of phospholipases that might be involved in exocytosis remains an open question.

The Monomeric G Protein Rab3A

It has become clear that the molecular basis of regulated exocytosis is conserved from yeast to mammalian cells. The process thus shares many proteins also involved in intracellular membrane fusion and constitutive secretion [11–15]. The Rab proteins are monomeric G proteins (molecular mass 20–30 kDa) and are the mammalian homologues of GTP-binding proteins that control vesicular trafficking in yeast [12,14]. Findings from the yeast system have had great impact on the study of exocytosis in mammalian cells. More than 30 different Rab proteins have been identified. They are implicated in the regulation of vectorial movement of vesicles to target membranes [12]. Although the precise mode of action of Rabs is not known, they may be involved in the assembly of the protein complex that regulates membrane fusion. Different Rab proteins seem to control distinct steps both in the biosynthetic/secretory and endocytotic pathways [12,14].

Rab3A is expressed almost exclusively in neurons and neuroendocrine cells. The protein is associated with synaptic vesicles from which it dissociates during or after the fusion of the vesicles with the presynaptic membrane [14]. We have demonstrated the presence of Rab3A on insulin-containing granules both by subcellular fractionation and by immuno-gold staining at the ultrastructural level [59]. As only a small fraction of the granules is released during stimulation of exocytosis in insulin-secreting cells, it was not possible to detect redistribution of Rab3A during the process. Initial results suggesting that the GTPases may influence exocytosis came from the use of synthetic peptides corresponding to the putative effector domain of Rab3A. When permeabilized cells were treated with such a peptide both basal and Ca^{2+}-stimulated insulin secretion were augmented [60]. Although the effect was not seen with peptides corresponding to the effector domain of Rab1 or the Ras protein, the use of such peptides only provides indirect evidence for the role of Rab3. Moreover, the four known isoforms of Rab3 (A, B, C and D) are identical in their effector domains and the peptides can therefore not be used to probe for the involvement of a given isoform. Hence, we chose an approach that permits the selective overexpression of the various isoforms and of mutated analogues with altered function. This approach has been used successfully in chromaffin cells and the derived cell line PC-12 [61,62]. The transient overexpression of wild type Rab3A in HIT-T15 cells did not alter insulin secretion. However, secretion both from intact and SLO-permeabilized cells was markedly inhibited by transfection with Rab3A mutants deficient either in GTPase activity or in the capacity of binding guanine nucleotides [59]. The effect is specific for Rab3, as the overexpression of Rab5, which controls endosome fusion, or its mutants, was ineffective [59]. These results were obtained by cotransfection of the hamster cell line with the plasmid under study and with cDNA encoding human proinsulin. This is a convenient means of assessing secretion from the 10%–15% of cells that take up and express the transgenes, as the assay of human C-peptide monitors secretion exclusively from the transfected cells [40]. The observation that functionally incompetent Rab3A protein inhibits exocytosis in insulin-secreting cells, and the findings that the protein is localized on secretory granules, point to an important role for the GTPase in insulin secretion. The results also emphasize the similarities between chromaffin cells and β cells with regard to the components of the exocytosis machinery.

Rab3B, Rab3C and Rab3D are all expressed in pancreatic islets and may also be implicated in insulin secretion [59]. Neither their subcellular localization nor their function is known at present. In native pancreatic β-cells the addition of GTP was found to stimulate slightly insulin exocytosis [63], while its poorly hydrolysable analogues GTPγS and GppNHp stimulate exocytosis in both native β-cells and derived cell lines [39,60,64]. It is unlikely that this effect is mediated by Rab3A [59] or, for that matter, by any monomeric GTPase. Whether the nucleotides act on a heterotrimeric G protein that would control exocytosis directly (independent of second messengers) remains to be established. The existence of such a "G_e" ("e" for exocytosis) has been inferred from experiments with GTP analogues in several cell systems, including mast cells and insulin secreting cells (for a discussion see [34]).

The results of the transfection experiments with Rab3A mutants strongly suggest that cycling of the G protein between a GDP- and a GTP-bound form is required for exocytosis to proceed. The functions of Rab proteins to undergo guanine nucleotide binding, to display intrinsic GTPase activity and to associate reversibly with membranes are controlled by several regulatory proteins (Fig. 5). In its GDP-liganded state Rab3A is found in the cytosol complexed with Rab-GDP-dissociation inhibitor (RabGDI). In HIT-T15 cells at least 50% of total Rab3A is recovered from the cytosol complexed to RabGDI [65]. The GDI acts as a chaperone for Rab3, targeting it to the secretory granule membrane [66]. Here another regulatory protein, the guanine

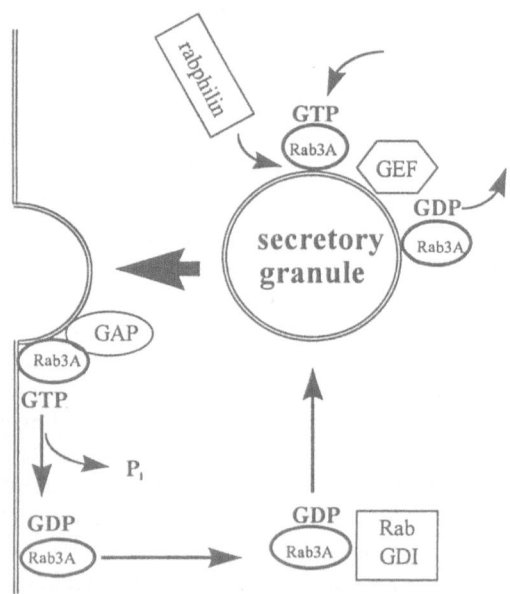

Fig. 5. The GDP/GTP cycle of Rab3 proteins and the putative role in insulin exocytosis of proteins interacting with Rab3. The binding of the guanine nucleotides to Rab3 as well as its membrane association are controlled by several regulatory proteins: *RabGDI*, Rab-GDP-dissociation inhibitor; *GEF*, guanine nucleotide exchange factor; and *GAP*, GTPase-activating protein. It is thought that the GTP-bound form of Rab3 proteins attracts rabphilin to the secretory granule (see text). Rabphilin is a Ca^{2+}/phospholipid-binding protein that may act as an effector in exocytosis

nucleotide exchange factor (GEF), stimulates the exchange of GDP for GTP (Fig. 5). Somehow the GTP-bound form of Rab3A favours the further progression of the exocytosis event, perhaps by attracting another protein, rabphilin, to the granule [67]. Neither GEF nor the subsequent regulatory protein GTPase activing protein (GAP) has so far been identified (Fig. 5). GAP would stimulate the relatively weak intrinsic GTPase activity of Rab3A during or after fusion of the granule with the plasma membrane [14]. The hydrolysis of GTP reforms the GDP-liganded Rab3 protein which is extracted from the membrane by RabGDI [65] to permit another cycle. As overexpression of Rab3A mutants incapable of cycling between a GDP and a GTP configuration inhibits stimulated insulin exocytosis [59], it is likely that the regulatory proteins controlling the guanine nucleotide cycle of Rab3A are targets for second messengers involved in stimulus-secretion coupling.

Second messengers could evoke both covalent and conformational alterations of the regulatory proteins. Phosphorylation/dephosphorylation could alter the function of the proteins. Although such modifications have not been demonstrated in secretory cells, it is of interest that overexpression of RabGDI in HIT-T15 cells inhibits nutrient-stimulated insulin secretion [68]. This finding suggests that there is a fine balance between Rab3A and its regulatory proteins which determines the rate of exocytosis. Ca^{2+}-binding proteins are of obvious interest as possible regulators of exocytosis. Rabphilin, first identified in brain, possesses two C2 domains in its C-terminal portion [67]. These are homologous to the C2 domain of protein kinase C and are also present in synaptotagmin [69] (more of the latter protein later). The C2 domains confer Ca^{2+} and phospholipid binding capacity to the proteins. Rab3A and its isoforms bind to the N-terminal portion of rabphilin in their GTP configuration [67] (Fig. 5). In chromaffin cells, overexpression of full-length rabphilin, but not of deletion mutants lacking one or both of the C2 domains, enhanced catecholamine secretion [70]. In view of the inhibition of exocytosis by the Rab3A mutant favouring the GTP-bound form, thereby attracting rabphilin to the vesicle, it has been proposed that Rab3A needs to dissociate from rabphilin to be permissive to exocytosis [70]. At present, rabphilin has not been detected in islets, but only in the insulin-secreting cell lines HIT-T15 and MIN6. The β-cell may contain a closely related protein not detected by analysis with probes for rabphilin mRNA or protein [59,71]. Nonetheless, we found, in agreement with the results in chromaffin cells, that overexpression of full-length rabphilin enhances insulin secretion from HIT-T15 cells (R. Regazzi and C.B. Wollheim, unpublished observations). Therefore, rabphilin or a homologue remains an attractive candidate protein as a target for Ca^{2+} action.

Many monomeric G proteins undergo posttranslational modifications at their C-terminal end. Through isoprenylation a geranyl-geranyl group is introduced into Rab3A and other low molecular weight GTPases including CDC42. This irreversible reaction renders the proteins more hydrophobic, which favours protein–protein interactions and membrane binding. It is noteworthy that certain prenylcysteine analogues, mimicking the C terminus of Rab3A, enhance exocytosis of insulin in permeabilized cells [72]. In contrast to isoprenylation, carboxymethylation is reversible and has therefore been suggested to represent a regulated step in cell activation [72,73]. CDC42 has been shown to be carboxymethylated in a GTP-dependent manner during glucose stimulation of pancreatic islets. This is not an obligatory event in exocytosis, as depolarizing K$^+$ concentrations did not cause CDC42 methylation [73].

The precise role of CDC42 in insulin secretion remains to be defined, but it could be involved in the regulation of the cytoskeleton and be implicated in the recruitment of granules during the second phase of insulin secretion [9].

The Components of the Exocytotic Machinery

The SNARE Hypothesis

The final steps of exocytosis are operationally subdivided into docking of vesicles to the plasma membrane, priming of the docked vesicles and ultimately fusion of the vesicles and plasma membranes (Fig. 6). Studies performed mainly in chromaffin cells have shown that priming requires ATP, but modifications of phospholipids on the secretory granules [74,75] and N-ethylmaleimide-sensitive factor (NSF) may also be involved [13,15]. Neither docking nor priming has been studied in detail in the β-cell. ATP depletion in SLO-permeabilized cells leads to about 90% attenuation of Ca^{2+}-induced insulin secretion, suggesting the existence of an ATP-independent component of exocytosis [40]. In neurosecretion, the vesicles dock in active zones at the presynaptic membrane, in close apposition to Ca^{2+} channels of the N-type [11,76]. Co-localization of insulin secretory granules and L-type Ca^{2+} channels has been observed in polarized β-cells [38]. The biphasic nature of insulin secretion has been taken as an indication of two pools of secretory granules, a readily releasable one and a larger, more slowly mobilizable one. The former may correspond to docked and/or primed

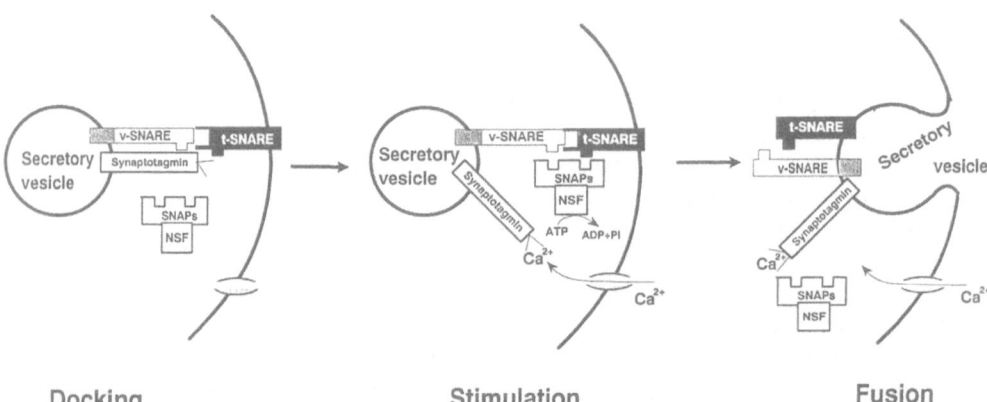

Docking   Stimulation   Fusion

Fig. 6. The SNARE hypothesis for regulated exocytosis. Secretory vesicles or granules are docked to the plasma membrane by the binding of vesicle-soluble NSF attachment receptor (*v-SNARE*) proteins to their cognite target membrane via *t-SNAREs*. The vesicular protein synaptotagmin is thought to clamp the vesicles in the docked state by blocking the access of the cytosolic soluble *N*-ethyl-maleimide-sensitive factor (*NSF*) attachment proteins (*SNAPs*) and of NSF to the SNAP receptor on the t-SNARE. During stimulation $[Ca^{2+}]_i$ rises and Ca^{2+} binds to the Ca^{2+}/phospholipid-binding protein synaptotagmin which permits the association with the SNAREs of SNAP and NSF, leading to the formation of the fusion complex. NSF-catalysed ATP hydrolysis may allow the fusion between the granular and plasma membranes to proceed

granules. The recruitment of the latter seems to require actin-containing microfilaments, as suggested from experiments with cytochalasins and *C.botulinum* C2 exotoxin [9]. However, the cytoskeleton has also been shown to act as a barrier, limiting the access of the granules to the plasma membrane [8], but, in contrast to chromaffin cells, secretagogues have not been shown to promote breakdown of the actin filamentous ring beneath the plasma membrane in the β-cell [8,9]. After fusion of the membranes a putative fusion pore is formed which spans the granular and plasma membranes [41]. Its regulation and molecular composition remain to be defined.

Remarkable progress has been made in recent years in our understanding of the molecular mechanisms controlling exocytosis. These studies have revealed that the proteins implicated in targeting and fusion of secretory vesicles with the plasma membrane are conserved from yeast to man [11–14,69]. The information was gained from (1) reconstitution of vesicle fusion in cell-free systems, (2) studies on mutant yeast strains incompetent for secretion, and (3) the use of clostridial neurotoxins. Rothman and colleagues [13] described several cytosolic proteins required for vesicle transport between Golgi compartments in a cell-free assay. This led to the discovery of NSF and soluble NSF attachment proteins (SNAPs; see Table 1). NSF is required for the fusion of a number of intracellular membranes. NSF and SNAPs form an ATP-dependent complex with membrane proteins. The same investigators used this approach to identify three brain proteins called SNAP receptors (SNAREs) [15]. These proteins were already known constituents of nerve terminals. They were the vesicle-associated membrane protein (VAMP) also called synaptobrevin, syntaxin, and synaptosomal-associated protein of 25 kDa (SNAP-25). The latter two are plasma membrane proteins (Table 1). According to the SNARE hypothesis every secretory vesicle bears a vesicular (v) SNARE, ensuring its targeting to the appropriate acceptor or target (t) membrane [13,15]. In vitro, v- and t-SNAREs form a 7S complex which associates with NSF/SNAP to form a 20S particle. The latter is dissociated in the

Table 1. Proteins regulating docking and fusion of secretory vesicles that are expressed in insulin-secreting cells

Protein	Location	Neurotoxin cleavage
VAMP-2/synaptobrevin-2	Secretory vesicle	tetanus, botulinum B, D, F
Cellubrevin	Secretory vesicle	tetanus, botulinum B, D, F
Synaptotagmin	Secretory vesicle	–
SNAP-25	Plasma membrane	botulinum A, E
Syntaxin	Plasma membrane	botulinum C
αSNAP	Cytosol	–
NSF	Cytosol	–
N-sec1	Plasma membrane	–
Rab3	Secretory vesicle/cytosol	–
Rabphilin	Secretory vesicle?	–
G$_i$/G$_o$	Plasma membrane	
	Secretory vesicle?	–

VAMP, vesicle-associated membrane protein; SNAP-25, synaptosomal-associated protein of 25 kDa; αSNAP, soluble NSF attachment protein; NSF, *N*-ethylmaleimide-sensitive factor; G$_i$/G$_o$, pertussin-toxin-sensitive heterotrimeric G proteins.

presence of ATP which is hydrolysed by NSF. The fusion particle can be recovered from brain extracts. These components of the fusion machinery are conserved throughout evolution, prompting the formulation of the SNARE hypothesis as a general model for fusion [15]. As illustrated in Fig. 6, αSNAP and NSF are recruited to the fusion site upon pairing of v-SNAREs with their cognite t-SNAREs. This pairing accomplishes vesicular docking to the proper fusion site. Confirmation of the key role of the various v- and t-SNAREs came from application of clostridial neurotoxins. The toxins were shown to cleave these proteins in a Zn^{2+}-dependent manner with high selectivity and high potency (Table 1) [77,78].

Application of the SNARE Hypothesis to Insulin Secretion

The v-SNAREs VAMP-2 and Cellubrevin

Of the three known v-SNAREs, VAMP-1, VAMP-2 and cellubrevin, the two latter were found in the β-cell and derived cell lines [64,79,80]. Cloning from an islet library revealed them to be identical to VAMP-2 and cellubrevin from brain [64]. As we could show by different methods, they are localized both on the secretory granules and on the GABA-containing synaptic-like vesicles [64]. Their role in insulin exocytosis was documented by the use of tetanus toxin and *C.botulinum* B toxin. These toxins were added to SLO-permeabilized cells because insulin-secreting cells, at variance with neuronal and neuroendocrine cells, do not express surface receptors for the clostridial neurotoxins permitting membrane permeation [77]. The cleavage of VAMP-2 and cellubrevin correlated extremely well with the inhibition of Ca^{2+}-stimulated insulin exocytosis over a range of tetanus toxin concentrations [64]. In contrast, Ca^{2+}-independent exocytosis evoked by GTPγS was unaffected by treatment with the neurotoxins [64]. This suggests that insulin-secreting cells may have two pathways for exocytosis, one of which does not involve VAMP-2 and cellubrevin. Alternatively, the action of GTPγS may be exerted on already docked vesicles, i.e., beyond the requirement for v-SNAREs. The elucidation of this intriguing result, which also may help in identifying the target of GTPγS, remains to be accomplished.

The t-SNAREs SNAP-25 and Syntaxin

The t-SNARE SNAP-25 has been localized to the plasma membrane in the β-cell [79-81]. Its functional role in insulin exocytosis was revealed by treatment of permeabilized HIT-T15 cells and native β-cells with *C.botulinum* A and E neurotoxins (Table 1). The toxins caused marked attenuation of Ca^{2+}-stimulated insulin secretion [81]. The escape from complete inhibition, contrasting with the effects of v-SNARE cleavage, could be explained by the in vitro observation that cleaved SNAP-25 can to some extent participate in SNARE complex formation.

Several isoforms of the t-SNARE syntaxin have been described, of which some have been found in the β-cell and β-cell lines [79,80]. Syntaxin-1 has been shown to be implicated in neurotransmitter exocytosis [11,69,76]. It is predominantly expressed on the β-cell plasma membrane [79,80]. In permeabilized islets a monoclonal antibody against syntaxin-1 was found to attenuate Ca^{2+}-stimulated insulin secretion [82]. Additional evidence for a key role for syntaxin was obtained from experiments with *C. botulinum* neurotoxin C1. This toxin abolished Ca^{2+}-induced insulin exocytosis both in permeabilized β-cells and HIT-T15 cells. In the former case only syntaxin-1 cleavage was observed, whereas the toxin caused an additional, minor cleavage of SNAP-25 in the cell line (J. Lang and C.B. Wollheim, unpublished observations). Inhibition of insulin secretion by stable overexpression of syntaxin-1A, but not of 1B, in βTC3 cells is difficult to reconcile with the aforementioned results. It is most likely that marked overexpression of one of the SNARE proteins reduces the availability of finely balanced factors for exocytosis. Therefore, caution must be taken always to verify the expression level and whenever possible overexpression should be checked by comparing the function of wild type and mutated proteins [59,61,62]. In particular the results with the C1 neurotoxin strongly suggest that intact syntaxin is a prerequisite for the exocytosis of insulin. Whether syntaxin functions to organize fusion by binding to Ca^{2+} channels [76], as has been found in neuronal cells, remains to be established.

nSec-1, whose function may be to stabilize syntaxin and to prevent its association with the v-SNAREs, is also expressed in insulin-secreting cells [80,83]. There is no information available on its role in insulin exocytosis.

The Ca²⁺-Binding Protein Synaptotagmin

Synaptotagmin contains two C2 domains which, like those of rabphilin [67], are homologous to the second conserved region of protein kinase C. The C2 domains confer Ca^{2+}- and phospolipid-binding properties to the protein. Synaptotagmin has been shown to bind in a Ca^{2+}-dependent manner to syntaxin [69]. In this way it may couple the synaptic vesicle to the Ca^{2+} entry site [69,76]. The role as Ca^{2+} sensor in neurotransmitter release has been attributed to synaptotagmin based in particular on findings in *Drosophila* mutants and in synaptotagmin I "knock-out" mice [69,84]. Nine isoforms of synaptotagmin are now known. Of these only synaptotagmin III and IV seem to be expressed in native β-cells [80,85,86] (J. Lang and C.B. Wollheim, unpublished observations). Insulin-secreting cell lines, on the other hand, express synaptotagmin I and II [86]. In these cells the isoforms are localized on insulin containing granules, while the subcellular distribution of the isoforms expressed in native β-cells has not yet been established. The Ca^{2+} affinity of synaptotagmin for phospholipid binding and in part also for syntaxin binding falls into the range of 2–$10\,\mu M$, a range compatible with $[Ca^{2+}]_i$ changes during insulin exocytosis. We have gathered compelling evidence for the crucial role of synaptotagmin in insulin exocytosis. Antibodies and Fab fragments directed against the first C2 domain of synaptotagmin I or II inhibited insulin exocytosis stimulated by Ca^{2+}, but not that evoked by GTPγS, in permeabilized INS-1 and HIT-T15 cells. Moreover, transient transfection of the latter cells with a mutant of synaptotagmin deficient in Ca^{2+}-dependent phospholipid binding attenuated K^+-stimulated insulin secretion in intact cells. The overexpression of wild-type synaptotagmin I and II had no effect [86]. These results support a role for synaptotagmin in insulin secretion evoked by glucose and other secretagogues that raise $[Ca^{2+}]_i$ in the β-cell. In neurosecretion synaptotagmin has been suggested to function as a clamping molecule (Fig. 6), removal of which following Ca^{2+} binding permits αSNAP and NSF to gain access to syntaxin and to other components in the SNARE complex. Whether this is indeed the mode of action of synaptotagmin is still debated [34].

The Cytosolic Proteins NSF and αSNAP

Insulin-secreting cells contain both the ATPase NSF and αSNAP and possibly βSNAP [87]. We examined the function of NSF and αSNAP in insulin secretion by adding the recombinant proteins to permeabilized cells. SLO-permeabilized cells display a run-down of Ca^{2+}-triggered insulin secretion with time. This is accompanied by marked losses of NSF and αSNAP from the cells. The run-down could be counteracted by addition of brain cytosol. Since N-ethylmaleimide (NEM) treatment of the cytosol prior to reconstitution attenuated the effect of cytosol, it appears that NSF and per-haps other NEM-sensitive proteins mediate the restoration of exocytosis. Addition of recombinant αSNAP, but not of NSF, also caused partial reconstitution. This demon-strates the implication of αSNAP in insulin exocytosis. With regard to NSF, it is likely that the 30%–40% of the protein remaining after the run-down period may suffice to support secretion, in particular if NSF is already attached to its protein partners [87]. It remains to be established at which steps in exocytosis NSF and αSNAP act. The prevailing idea is that the proteins act after the pairing of the SNAREs and after Ca^{2+} binding to synaptotagmin (Fig. 6). In this scenario they would initiate the fusion process itself. Another possibility is that NSF functions as a chaperone conferring fusion competence to the granule. Such a prefusion role of NSF may explain why αSNAP and NSF have also been found on undocked synaptic vesicles [88]. Other molecular chaperones may well be involved in the ordered assembly and folding of the various proteins that mediate docking and fusion.

Conclusion and Perspectives

Investigation in the 1960s of the mechanism of insulin secretion permitted to recog-nize the overall pattern of the jigsaw puzzle [1]. Over the past 25 years the fine details of the picture have slowly emerged through application of new techniques of cellular and molecular biology. The complexity of the process of stimulus-secretion coupling in the β-cell has not been reduced through increased knowledge, but the puzzle now contains well-defined holes that most certainly will be filled with ever increasing speed. The understanding of the molecular basis of vesicular fusion from yeast to mammalian neurons has helped to define a number of proteins on the insulin secre-tory granule and at the β-cell membrane that participate in the exocytosis of insulin. It is hoped that the fusion protein itself will be discovered soon, permitting the completion of this part of the β-cell puzzle. The methods allowing the study in situ of protein-protein interactions and the point mutation of the critical domains of the interacting sequences will most certainly allow us to pinpoint the targets for Ca^{2+}, protein kinases and G proteins, solving questions already posed by Albert Renold nearly three decades ago [1]. The intimate link between the mitochondria and the regulation of exocytosis represents another unfinished part of the puzzle. The further characterization of metabolism-secretion coupling may well reveal yet another facet of the complexity of the insulin-producing cell.

Acknowledgements. The original work referred to in this article was performed by many collaborators to whom I am deeply indebted. I should particularly like to thank Dr. Susanne Ullrich, Dr. William Pralong, Dr. Romano Regazzi and Dr. Jochen Lang without whose innovative ideas, hard work and never failing enthusiasm the projects on stimulus-secretion coupling in the β-cell would not have been successful. I am also grateful to Mrs. Jean Gunn for invaluable help with the editing of the manuscript and to Mrs. Mona Guichoud for expert secretarial assistance. The continuous support of our work by the Swiss National Science Foundation is gratefully acknowledged.

References

1. Renold AE (1970) Insulin biosynthesis and secretion – a still unsettled topic. N Engl J Med 282: 173–182
2. Porte D Jr (1991) β-cells in type II diabetes mellitus. Banting Lecture 1990. Diabetes 40: 166–180
3. Turner RC, Hattersley AT, Shaw JTE, Levy JC (1995) Type II diabetes: clinical aspects of molecular biological studies. Diabetes 44: 1–10
4. Froguel PH, Zouali H, Vionnet N, Velho G, Vaxillaire M, Sun F, Lesage S, Stroffel M, Takeda J, Passa PH, Permutt MA, Beckmann JS, Bell GI, Cohen D (1993) Familial hyperglycemia due to mutations in glucokinase: definition of a subtype of diabetes mellitus. N Engl J Med 328: 697–702
5. Maassen JA, Kadowaki T (1996) Maternally inherited diabetes and deafness: a new diabetes subtype. Diabetologia 39: 375–382
6. Orci L (1982) Macro- and micro-domains in the endocrine pancreas. Banting Lecture 1981. Diabetes 31: 538–565
7. Wollheim CB, Sharp GWG (1981) Regulation of insulin release by calcium. Physiol Rev 61: 914–973
8. Howell SL, Tyhurst M (1986) The cytoskeleton and insulin secretion. Diab Metab Rev 2: 107–123
9. Li G, Rungger-Brändle E, Just I, Jonas JC, Aktories K, Wollheim CB (1994) Effect of disruption of actin filaments by *Clostridium Botulinum* C2 toxin on insulin secretion in HIT-T15 cells and pancreatic islets. Molec Biol of the Cell 5: 1199–1213
10. Burgess TL, Kelly RB (1987) Constitutive and regulated secretion of proteins. Annu Rev Cell Biol 3: 243–293
11. Damer CK, Creutz CE (1994) Secretory and synaptic vesicle membrane proteins and their possible roles in regulated exocytosis. Progress in Neurobiology 43: 511–536
12. Schekman R, Orci L (1996) Coat proteins and vesicle budding. Science 271: 1526–1539
13. Rothman JE (1994) Mechanisms of intracellular protein transport. Nature 372: 55–63
14. Ferro-Novick S, Jahn R (1994) Vesicle fusion from yeast to man. Nature 370: 191–193
15. Söllner T, Whiteheart SW, Brunner M, Erdjument-Bromage H, Geromanos S, Tempst P, Rothman JE (1993) SNAP receptors implicated in vesicle targeting and fusion. Nature 362: 318–324
16. Boucher BJ, Mannan N, Noonan K, Hales CN, Evans SJW (1995) Glucose intolerance and impairment of insulin secretion in relation to vitamin D deficiency in East London Asians. Diabetologia 38: 1239–1245
17. Pozzan T, Rizzuto R, Volpe P, Meldolesi J (1994) Molecular and cellular physiology of intracellular calcium stores. Physiol Rev 74: 595–636
18. Ashcroft FM, Proks P, Smith PA, Ämmälä C, Bokvist K, Rorsman P (1994) Stimulus-secretion coupling in pancreatic β-cells. J Cell Biochem 55S: 54–65
19. Montero M, Brini M, Marsault R, Alvarez J, Sitia R, Pozzan T, Rizzuto R (1995) Monitoring dynamic changes in free Ca²⁺ concentration in the endoplasmic reticulum of intact cells. EMBO J 14: 5467–5475

20. Prentki M, Biden TJ, Janjic D, Irvine RF, Berridge MJ, Wollheim CB (1984) Rapid mobilization of Ca^{2+} from rat insulinoma microsomes by inositol-1,4,5-trisphosphate. Nature 309: 562–564

21. Wollheim CB, Biden TJ (1986) Signal transduction in insulin secretion: comparison between fuel stimuli and receptor agonists. Ann NY Acad Sci 488: 317–333

22. Biden TJ, Peter-Riesch B, Schlegel W, Wollheim CB (1987) Ca^{2+}-mediated generation of inositol 1,4,5-trisphosphate and inositol 1,3,4,5-tetrakisphosphate in pancreatic islets. J Biol Chem 262: 3567–3571

23. Ashcroft FM, Harrison DE, Ashcroft JH (1984) Glucose induces closure of single potassium channels in isolated rat pancreatic β-cells. Nature 312: 466–468

24. Inagaki N, Gonoi T, Clement IV JP, Namba N, Inazawa J, Gonzalez G, Aguilar-Bryan L, Seino S, Bryan J (1995) Reconstitution of I_{KATP}: An inward rectifier subunit plus the sulfonylurea receptor. Science 270: 1166–1170

25. Aguilar-Bryan L, Nichols CG, Wechsler SW, Clement IV JP, Boyd III AE, Gonzalez G, Herrera-Sosa H, Nguy K, Bryan J, Nelson DA (1995) Cloning of the β-cell high-affinity sulfonylurea receptor: A regulator of insulin secretion. Science 268: 423–426

26. Pralong WF, Bartley C, Wollheim CB (1990) Single islet β-cell stimulation by nutrients: relationship between pyridine nucleotides, cytosolic Ca^{2+} and secretion. EMBO J 9: 53–60

27. Pralong WF, Spät A, Wollheim CB (1994) Dynamic pacing of cell metabolism by intracellular Ca^{2+} transients. J Biol Chem 269: 27310–27314

28. Santos RM, Rosario LM, Nadal A, Garcia-Sancho J, Soria B, Valdeolmillos M (1991) Widespread synchronous $[Ca^{2+}]_i$ oscillations due to bursting electrical activity in single pancreatic islets. Pflügers Arch 418: 417–422

29. McCormack JG, Halestrap AP, Denton RM (1990) Role of calcium ions in regulation of mammalian intramitochondrial metabolism. Physiol Rev 70: 391–425

30. Kennedy ED, Rizzuto R, Theler J-M, Pralong W-F, Bastianutto C, Pozzan T, Wollheim CB (1996) Glucose-stimulated insulin secretion correlates with changes in mitochondrial and cytosolic Ca^{2+} in aequorin-expressing INS-1 cells. J Clin Invest 98: 2524–2538

31. Gembal M, Detimary P, Gilon P, Gao ZY, Henquin JC (1993) Mechanisms by which glucose can control insulin release independently from its action on adenosine triphosphate-sensitive K^+ channels in mouse β-cells. J Clin Invest 91: 871–880

32. Gilon P, Shepherd RM, Henquin JC (1993) Oscillations of secretion driven by oscillations of cytoplasmic Ca^{2+} as evidenced in single pancreatic islets. J Biol Chem 268: 22265–22268

33. Hellman B, Gylfe E, Bergsten P, Grapengiesser E, Lund PE, Berts A, Tengholm A, Pipeleers DG, Ling Z (1994) Glucose induces oscillatory Ca^{2+} signalling and insulin release in human pancreatic β-cells. Diabetologia 37 (suppl 2): S11–S20

34. Wollheim CB, Lang J, Regazzi R (1996) The exocytotic process of insulin secretion and its regulation by Ca^{2+} and G-proteins. Diabetes Rev 4: 276–297

35. Ämmälä C, Eliasson L, Bokvist K, Larsson O, Ashcroft FM, Rorsman P (1993) Exocytosis elicited by action potentials and voltage-clamp calcium currents in individual mouse pancreatic β-cells. J Physiol 472: 665–688

36. Yaseen A, Pedley KC, Howell SL (1992) Regulation of insulin secretion from islets of Langerhans rendered permeable by electric discharge. Biochem J 206: 81–87

37. Ullrich S, Prentki M, Wollheim CB (1990) Somatostatin inhibition of Ca^{2+}-induced insulin in permeabilized HIT-T15 cells. Biochem J 270: 273–276

38. Bokvist K, Eliasson L, Ämmälä C, Renstrom E, Rorsman P (1995) Co-localization of L-type Ca^{2+} channels and insulin-containing secretory granules and its significance for the initiation of exocytosis in mouse pancreatic β-cells. Embo J 14: 50–57

39. Vallar L, Biden TJ, Wollheim CB (1987) Guanine nucleotides induce Ca^{2+}-independent insulin secretion from permeabilized RINm5F cells. J Biol Chem 262: 5049–5056

40. Lang J, Nishimoto I, Okamoto T, Regazzi R, Kiraly C, Weller U, Wollheim CB (1995) Direct control of exocytosis by receptor-mediated activation of the heterotrimeric GTPases G_i and G_o or by the expression of their active $G\alpha$ subunits. EMBO J 14: 3635–3644

41. Parsons TD, Coorssen JR, Horstmann H, Almers W (1995) Docked granules, the exocytic burst, and the need for ATP hydrolysis in endocrine cells. Neuron 15: 1085–1096

42. Theler JM, Mollard P, Guérineau N, Vacher P, Pralong WF, Schlegel W, Wollheim CB (1992) Video imaging of cytosolic Ca^{2+} in pancreatic β-cells stimulated by glucose, carbachol and ATP. J Biol Chem 267: 18110–18117

43. Thorens B, Waeber G (1993) Glucagon-like peptide-I and the control of insulin secretion in the normal state and in NIDDM. Diabetes 42: 1219–1225

44. Ämmälä C, Ashcroft FM, Rorsman P (1993) Calcium-independent protentiation of insulin release by cyclic AMP in single-β-cells. Nature 363: 356–358

45. Holz IV GG, Kühtreiber WM, Habener JF (1993) Pancreatic β-cells are rendered glucose-competent by the insulinotropic hormone glucagon-like peptide-1 (7–37). Nature 361: 362–365

46. Jones PM, Persaud SJ, Howell SL (1992) Ca²⁺-induced insulin secretion from electrically permeabilized islets. Loss of the Ca²⁺-induced secretory response is accompanied by loss of Ca²⁺-induced protein phosphorylation. Biochem J 285: 973–978

47. Baffy G, Yang L, Wolf BA, Williamson JR (1993) G-protein specificity in signalling pathways that mobilize calcium in insulin-secreting β-TC3 cells. Diabetes 42: 1878–1882

48. Wollheim CB, Regazzi R (1990) Protein kinase C in insulin releasing cells: putative role in stimulus secretion coupling. FEBS Lett 268: 376–380

49. Ämmälä C, Eliasson L, Bokvist K, Berggren P-O, Honkanen RE, Sjöholm A, Rorsman P (1994) Activation of protein kinases and inhibition of protein phosphatases play a central role in the regulation of exocytosis in mouse pancreatic β-cells. Proc Natl Acad Sci USA 91: 4343–4347

50. Wenham RM, Landt M, Easom RA (1994) Glucose activates the multifunctional Ca²⁺/calmodulin-dependent protein kinase II in isolated rat pancreatic islets. J Biol Chem 269: 4947–4952

51. Ribar TJ, Epstein PN, Overbeek PA, Means AR (1995) Targeted overexpression of an inactive calmodulin that binds Ca²⁺ to the mouse pancreatic β-cell results in impaired secretion and chronic hyperglycemia. Endocrinology 136: 106–115

52. Rodbell M (1992) The role of GTP-binding proteins in signal transduction: from the sublimely simple to the conceptually complex. Curr Top Cell Regul 32: 1–47

53. Neer EJ (1995) Heterotrimeric G proteins: organizers of transmembrane signals. Cell 80: 249–257

54. Zigman JM, Westermark GT, LaMendola J, Steiner DF (1994) Expression of cone transducin, Gz α, and other G-protein α-subunit messenger ribonucleic acids in pancreatic islets. Endocrinology 135: 31–37

55. Ullrich S, Wollheim CB (1988) GTP-dependent inhibition of insulin secretion by epinephrine in permeabilized RINm5F cells. J Biol Chem 263: 8615–8620

56. Ullrich S, Wollheim CB (1989) Galanin inhibits insulin secretion by direct interference with exocytosis. FEBS Lett 247:401–404

57. Nilsson T, Arkhammar P, Rorsman P, Berggren P-O (1988) Inhibition of glucose-stimulated insulin release by α_2-adrenoceptor activation is paralleled by both a repolarization and a reduction in cytoplasmic free Ca²⁺ concentration. J Biol Chem 263: 1855–1860

58. Tamagawa T, Niki I, Niki H, Niki A (1985) Catecholamines inhibit insulin release independently of changes in cytosolic free Ca²⁺. Biomed Res 6: 429–432

59. Regazzi R, Ravazzola M, Iezzi M, Lang J, Zahraoui A, Andereggen E, Morel P, Takai Y, Wollheim CB (1996) Expression, localization and functional role of small GTPases of the Rab3 family in insulin-secreting cells. J Cell Sci 109: 2265–2273

60. Li G, Regazzi R, Balch WE, Wollheim CB (1993) Stimulation of insulin release from permeabilized HIT-T15 cells by a synthetic peptide corresponding to the effector domain of the small GTP-binding protein rab3. FEBS Lett 327: 145–149

61. Holz R, Brondyk WH, Senter RA, Kuzion L, Macara IG (1994) Evidence for the involvement of Rab3A in Ca²⁺-dependent exocytosis from adrenal chromaffin cells. J Biol Chem 269: 10229–10234

62. Johannes L, Lledo P-M, Roa M, Vincent J-D, Henry J-P, Darchen F (1994) The GTPase Rab3a negatively controls calcium-dependent exocytosis in neuroendocrine cells. EMBO J 13: 2029–2037

63. Wollheim CB, Ullrich S, Meda P, Vallar L (1987) Regulation of exocytosis in electrically permeabilized insulin-secreting cells. Evidence for Ca²⁺ dependent and independent secretion. Biosci Rep 7: 443–454

64. Regazzi R, Wollheim CB, Lang J, Theler JM, Rossetto O, Montecucco C, Sadoul K, Weller U, Palmer M, Thorens B (1995) VAMP-2 and cellubrevin are expressed in pancreatic beta-cells and are essential for Ca²⁺ – but not for GTP gamma-s-induced insulin secretion. EMBO J 14: 2723–2730

65. Regazzi R, Kikuchi A, Takai Y, Wollheim CB (1992) The small GTP-binding proteins in the cytosol of insulin-secreting cells are complexed to GDP dissociation inhibitor proteins. J Biol Chem 267: 17512–17519

66. Pfeffer SR, Dirac-Svejstrup AB, Soldati T (1995) Rab GDP dissociation inhibitor: putting Rab GTPases in the right place. J Biol Chem 270: 17057–17059

67. Li C, Takei K, Geppert M, Daniell L, Stenius K, Chapman ER, Jahn R, De Camilli P, Südhof TC (1994) Synaptic targeting of Rabphilin-3A, a synaptic vesicle Ca2+/phospholipid-binding protein, depends on rab3A/3C. Neuron 13: 885–898

68. Wollheim CB, Regazzi R, Iezzi M, Lang J (1996) Identification of proteins involved in the regulation of exocytosis of insulin in the pancreatic β-cell. Eur J Cell Biol 69 (suppl 42) A 407

69. Sudhof TC (1995) The synaptic vesicle cycle: a cascade of protein-protein interactions. Nature 375: 645–653

70. Chung S-H, Takai Y, Holz RW (1995) Evidence that the Rab3a-binding protein, Rabphilin3a, enhances regulated secretion. J Biol Chem 270: 16714–16718

71. Inagaki N, Mizuta M, Seino S (1994) Cloning of a mouse Rabphilin-3A expressed in hormone-secreting cells. J Biochem Tokyo 116: 239–242

72. Regazzi R, Sasaki T, Takahashi K, Jonas JC, Volker C, Stock JB, Takai Y, Wollheim CB (1995) Prenylcysteine analogs mimicking the C-terminus of GTP-binding proteins stimulate exocytosis from permeabilized HIT-T15 cells. Biochim Biophys Acta 1268: 268–278

73. Kowluru A, Seavey SE, Li G, Sorenson RL, Weinhaus AJ, Nesher R, Rabaglia ME, Vadakekalam J, Metz SA (1996) Glucose- and GTP-dependent stimulation of the carboxyl methylation of CDC42 in rodent and human pancreatic islets and pure β-cells. J Clin Invest 98: 540–555

74. De Camilli P, Enr SD, McPherson PS, Novick P (1996) Phosphoinositides as regulators in membrane traffic. Science 271: 1533–1539

75. Hay JC, Fisette PL, Jenkins GH, Fukami K, Takenawa T, Anderson RA, Martin TF (1995) ATP-dependent inositide phosphorylation required for Ca²⁺-activated secretion. Nature 374: 173–177

76. Sheng ZH, Rettig L, Cook T, Catterall WA (1996) Calcium-dependent interaction of N-type calcium channels with the synaptic core complex. Nature 379: 451–454

77. Niemann H, Blasi J, Jahn R (1994) Clostridial toxins: new tools for dissecting exocytosis. Trends in Cell Biol 4: 179–185

78. Rossetto O, Deloye F, Poulain B, Pellizzari R, Schiavo G, Montecucco C (1995) The metalloproteinase activity of tetanus and botulism neurotoxins. J Physiol 89: 43–50

79. Jacobsson G, Bean AJ, Scheller RH, Juntti-Bergren L, Deeney JT, Berggren PO, Meister B (1994) Identification of synaptic proteins and their isoform mRNAs in compartments of pancreatic endocrine cells. Proc Natl Acad Sci USA 91: 12487–12491

80. Wheeler MB, Sheu L, Ghai M, Bouquillon A, Grondin G, Weller U, Beadoin AR, Bennet M, Trimble WS, Gaisano HY (1996) Characterization of SNARE protein expression in β cell lines and pancreatic islets. Endocrinology 137: 1340–1348

81. Sadoul K, Lang J, Montecucco C, Weller U, Regazzi R, Catsicas S, Wollheim CB, Halban PA (1995) SNAP-25 is expressed in islets of Langerhans and is involved in insulin release. J Cell Biol 128: 1019–1028

82. Martin F, Moya F, Gutierrez LM, Reig JA, Soria B (1995) Role of syntaxin in mouse pancreatic beta cells. Diabetologia 38: 860–863

83. Katagiri H, Terasaki J, Tomiyasu M, Ishihara H, Ogihara T, Inukai K, Fukushima Y, Anai M, Kikuchi M, Miyazaki J, Yazaki Y, Oka Y (1995) A novel isoform of syntaxin-binding protein homologous to yeast sec1 expressed ubiquitously in mammalian cells. J Biol Chem 270: 4963–4966

84. Littleton JT, Bellen HJ (1995) Synaptotagmin controls and modulates synaptic-vesicle fusion in a Ca²⁺-dependent manner. Trends Neurosci 18: 177–183

85. Mizuta M, Inagaki N, Nemoto Y, Matsukura S, Takahashi M, Seino S (1994) Synaptotagmin III is a novel isoform of rat synaptotagmin expressed in endocrine and neuronal cells. J Biol Chem 269: 11675–11678

86. Lang J, Fukuda M, Zhang H, Kiraly C, Mikoshiba K, Wollheim CB (1996) Subcellular localisation and role of synaptotagmin in insulin-secreting pancreatic β-cell lines. Diabetologia 39 (suppl 1) A115

87. Kiraly-Borri C, Morgan A, Burgoyne R, Weller U, Wollheim C, Lang J (1996) Soluble N-ethylmaleimide sensitive factor attachment protein and N-ethylmaleimide insensitive factors are required for Ca²⁺ stimulated exocytosis of insulin. Biochem J 314: 199–203

88. Morgan A, Burgoyne RD (1995) Is NSF a fusion protein? Trends Cell Biol 5: 335–339

The Mechanism of Insulin Receptor Binding, Activation and Signal Transduction

P. De Meyts and K. Seedorf

Introduction

The insulin receptor belongs to the family of receptor protein tyrosine kinases (RTK). It has a covalent, disulphide-bridged, dimeric ($\alpha 2\beta 2$) subunit structure, of which only the three-dimensional structure of the tyrosine kinase domain has so far been solved [1–3]. Upon insulin binding, the β subunits become phosphorylated on a well-defined set of tyrosine residues (Tyr_{960}, Tyr_{1146}, Tyr_{1150}, Tyr_{1151}, Tyr_{1316}, Tyr_{1322}), presumably through a transphosphorylation mechanism between the two kinase domains [1,4]. The mechanism whereby this activation occurs is presumably a variation on a theme common to the RTKs and cytokine receptor families, that is ligand-induced receptor crosslinking or dimerization. Recent work has indeed led to the suggestion that insulin behaves as a monomeric, but bivalent, ligand that crosslinks distinct subsites on the two receptor α subunits [5,6]. This review will focus first on the implications of the dimerizing/crosslinking mechanism of receptor activation, which we have discussed in several recent reviews [5,7–9]. Then, recent progress in the understanding of downstream signalling pathways will be described, also updating recent reviews [10–12]. Finally, we will describe how this new knowledge is being used to design new experimental genetic approaches to create polygenic animal models of non-insulin-dependent diabetes mellitus (NIDDM).

Implications of Receptor Activation by Ligand-Induced Dimerization or Crosslinking

The receptors of the tyrosine kinase and cytokine superfamilies are activated by a common mechanism, i.e. the homo- or hetero-dimerization of two receptor components upon ligand binding, or possibly sometimes oligomerization to a higher degree [9,13–17].

The initial concept envisaged either that the ligand would dimerize the receptors by inducing a conformational change that brings two monomeric ligand–receptor complexes together, resulting in a 2:2 stoichiometry (as initially proposed for epidermal growth factor, EGF), or that a dimeric ligand (e.g. platelet-derived growth factor, PDGF) would bring the two receptors together. In many cases it now appears, in fact, that a monomeric ligand featuring two distinct binding surfaces (that is, functionally bivalent) brings the two receptors together in a sequential fashion, resulting in a 1:2 stoichiometry. Such a mechanism operates in the cytokine receptor family, as first demonstrated by the crystallographic resolution of the growth hormone complex with the extracellular part of its receptor [17,18]. In some cases the two receptor halves are not identical but rather are distinct subunits, resulting in a heterodimeric complex. The dimerization of the receptors provides a simple mechanism for triggering downstream signalling by bringing together the two intra-cellular kinase domains of the receptor and allowing transphosphorylation to occur. In the cytokine family, the tyrosine kinase is not an intrinsic part of the receptor; rather, a cytoplasmic kinase (such as JAK or Fyn) is recruited to the receptor after activation [19].

Another special case exists in the insulin/insulin-like growth factor-I (IGF-I) receptor tyrosine kinase subfamily, where the receptors are covalent dimers even in the absence of ligand. The receptors feature two extracellular α subunits that contain the ligand-binding domains and two transmembrane β subunits that contain the tyrosine kinase domain. We and others have recently provided evidence based on structure–function relationships of insulin analogues and receptor-binding kinet-ics that a binding mechanism similar to that proposed for monomeric receptor

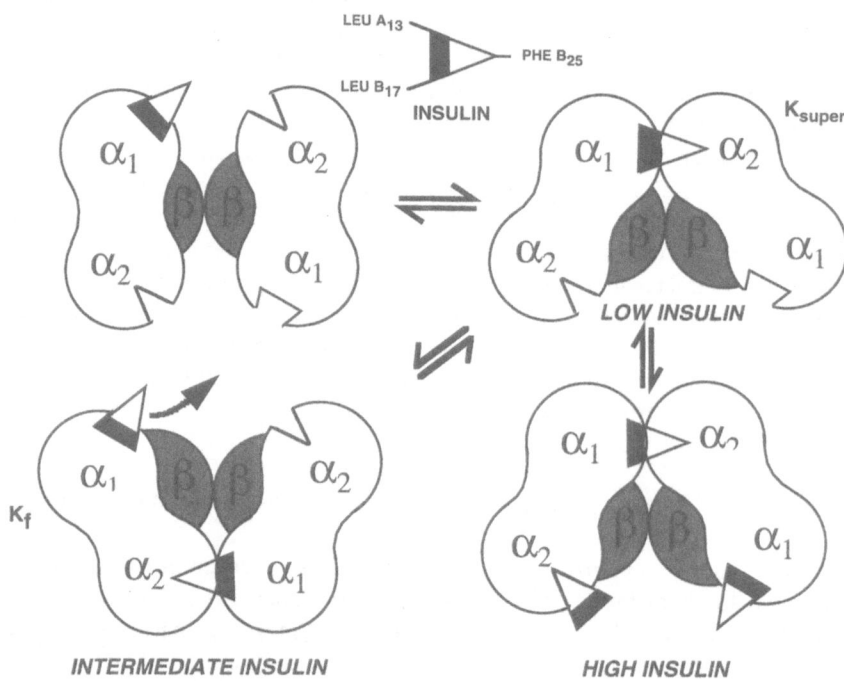

Fig. 1. The symmetrical, alternative crosslinking model for insulin binding to its receptor. The receptor is viewed from the top. Each α subunit is represented as containing two subsites $\alpha 1$ and $\alpha 2$. The first insulin molecule binds through its hexamer-forming surface to $\alpha 1$, and then crosslinks through its dimer-forming surface to $\alpha 2$ on the second α subunit. The resulting tight bivalent binding is referred to as the K_{super} state. If the concentration of insulin is increased, partial dissociation of the first bound insulin allows a second insulin molecule to crosslink the opposite $\alpha 1$–$\alpha 2$ pair, which allows the first molecule to dissociate completely. At very high insulin concentrations, $\alpha 1$ and $\alpha 2$ opposite the first cross-link are both occupied, preventing the second cross-linking and maintaining the first bound insulin molecule in the K_{super} state, explaining the bell-shaped dose-response curve of dissociation kinetics. (Adapted from [5], where a more complete explanation of the model can be found)

tyrosine kinases and cytokine receptors is operating [5,6]. This model postulates that the insulin molecule has two binding surfaces, one overlapping with the surface involved in insulin dimerization and the other involved in insulin hexamerization. Insulin thus behaves as a bivalent ligand which sequentially crosslinks two distinct binding domains, one on each α subunit of the insulin receptor dimer (Fig. 1).

The recent finding that IGF-I binding to its receptor has kinetic properties similar to those of insulin binding to its receptor suggests that the binding mechanism is probably similar [20,21].

The binding model in which a bivalent ligand induces sequential dimerization of its receptor (or in the case of insulin, crosslinking within a covalent dimeric structure) has a number of interesting properties as compared with classical "monovalent" binding models (e.g. bell-shaped biological dose–response curves [9,23]). Perhaps the most interesting aspect from a functional viewpoint is that dimerization or crosslinking provides both a switch and a timer for receptor activation right at the receptor-binding step [5,7–9]. The switch consists in the bivalent dimerization bringing together the two kinase domains. The timer is provided by the effect of dimerization on ligand dissociation kinetics. Indeed, the bivalently bound ligand is much less likely to dissociate from the receptor than the monovalently bound one, greatly enhancing the probability that the complex will be able to initiate sustained signalling.

In the case of the insulin and IGF-I receptors, an added feature regulates the "timer" component of the binding step: the existence of negative cooperativity in binding which modulates the ligand dissociation rate as a function of ambient ligand concentration (i.e. receptor occupancy) [5,24–26]. We have recently proposed that this phenomenon may result from the formation of a second crosslinking bridge between the two receptor α subunits by a second insulin molecule within a symmetrical receptor structure [5] (Fig. 1).

The slowing down of dissociation kinetics by bivalent crosslinking and its regulation by negative cooperativity appear to play an important role in determining the mitogenic potency of insulin and insulin analogues by generating sustained downstream signalling [27,28], underscoring the importance of the time factor in signalling specificity as discussed elsewhere [8].

Interestingly, recent data suggest that the paradigm of ligand-induced dimerization may not be limited to cytokine receptors and receptor tyrosine kinases, but may apply also to heptahelical, G-coupled receptors such as the β_2-adrenergic receptor [29]. Such a mechanism may finally explain old findings of apparent negative cooperativity and ligand-induced accelerated tracer dissociation in the β-adrenergic [30] and thyroid-stimulating hormone (TSH) receptors [31], difficult to reconcile with a monomeric stoichiometry. Furthermore, activation by dimerization is not restricted to membrane receptors, but operates in receptors of the steroid/retinoic acid family, signalling molecules such as the kinase Raf or the adaptor Grb2 (see below) and a variety of transcription factors and repressors. Therefore, ligand-induced dimerization now appears to be a widespread regulatory mechanism, besides other well-established mechanisms such as covalent modifications (e.g. phosphorylation), precursor or proenzyme processing, and allosteric regulation. The full implications of this paradigm are only beginning to be appreciated.

Downstream Signalling Pathways of the Insulin Receptor Tyrosine Kinase (Fig. 2)

The IRS Family

Unlike other receptor tyrosine kinases, the tyrosine phosphorylated insulin (and IGF-I) receptors do not associate strongly with downstream SH2 (src homology 2) domain-containing signalling molecules, but use "docking molecules" such as the insulin receptor substrates 1 and 2 (IRS-1/2), or Shc (for Src-homology 2/α-collagen) [10,32,33]. IRS-2 (also called 4PS) was recently cloned from myeloid progenitor cells [34] and compensates for the loss of IRS-1 in IRS-1 knock-out mice (see below) [35–37]. A 190 kDa homologue was cloned from simian COS cells that may be the simian IRS-1, but Southern blotting in Chinese hamster ovary (CHO) cells suggested the coexistence of IRS-1 and this Cos-related pp190 [38]. A novel Grb2-associated IRS-1-related docking protein with a predicted molecular mass of 77 kDa (Gab1) was cloned from a glial tumour [39]. Interestingly, it also associated with the EGF receptor, which thus far was not known to use docking proteins for signalling. A pp160 protein that crossreacts weakly with anti-IRS-1 and anti-IRS-2 antibodies and co-precipitates with insulin receptor, Shc and Grb2 was recently identified in an insulin growth-dependent T-cell lymphoma line (LB). The purification of this protein is in progress [40]. Thus, it appears that the IRS family of docking proteins may expand further.

Pleckstrin Homology and Phosphotyrosine Binding Domains

IRS-1/2 reacts with upstream signalling elements via two specialized domains in its N-terminal domain: IH1^{PH}, a pleckstrin homology (PH) domain, and IH2^{PTB}, a phosphotyrosine binding (PTB) domain [32]. PTB domains and PH domains appear to be variations on the same structural theme [41]. PH domains have been reported to bind phospholipids, G proteins β/γ subunits and/or protein kinase C (PKC) isoforms [42], but the specificity of some of these interactions has been questioned [41]. The main function of PH/PTB domains may be to recruit signalling molecules to the cell surface [41] and to couple phosphatidylinositol signalling to GTP hydrolysis. The crystal structure of some PH domains bound to inositol phosphates has been solved [43–45]. IH1^{PH} may help docking IRS-1 to the plasma membrane [46,47] or be involved in crosstalk with G-coupled receptors (see below). Surprisingly, however, Conway et al. recently reported that an IRS-1 lacking the first conserved segment of the PH domain was found in the low density microsome (LDM) fraction of transfected CHO cells to the same extent as full-length IRS-1, and they concluded that the PH domain was not necessary for membrane association [48]. The insulin-stimulated tyrosine phosphorylation of the mutant IRS-1 was, however, reduced or completely abrogated in vivo [46–48], but not in vitro [47]. The phosphorylation of the mutant IRS-1 could be rescued in vivo by overexpressing the insulin receptor [46]. Thus, it is clear that the PH domain is required for sensitive coupling of IRS-1 to insulin receptors.

IH2^{PTB} has now clearly been shown to bind to the insulin receptor juxtamembrane region comprising the phosphorylated Tyr$_{960}$ in a NPEY motif (LxxxxNPXYXSXSD) [32]. The crystal structure of the IH2^{PTB} domain alone and in complex with the

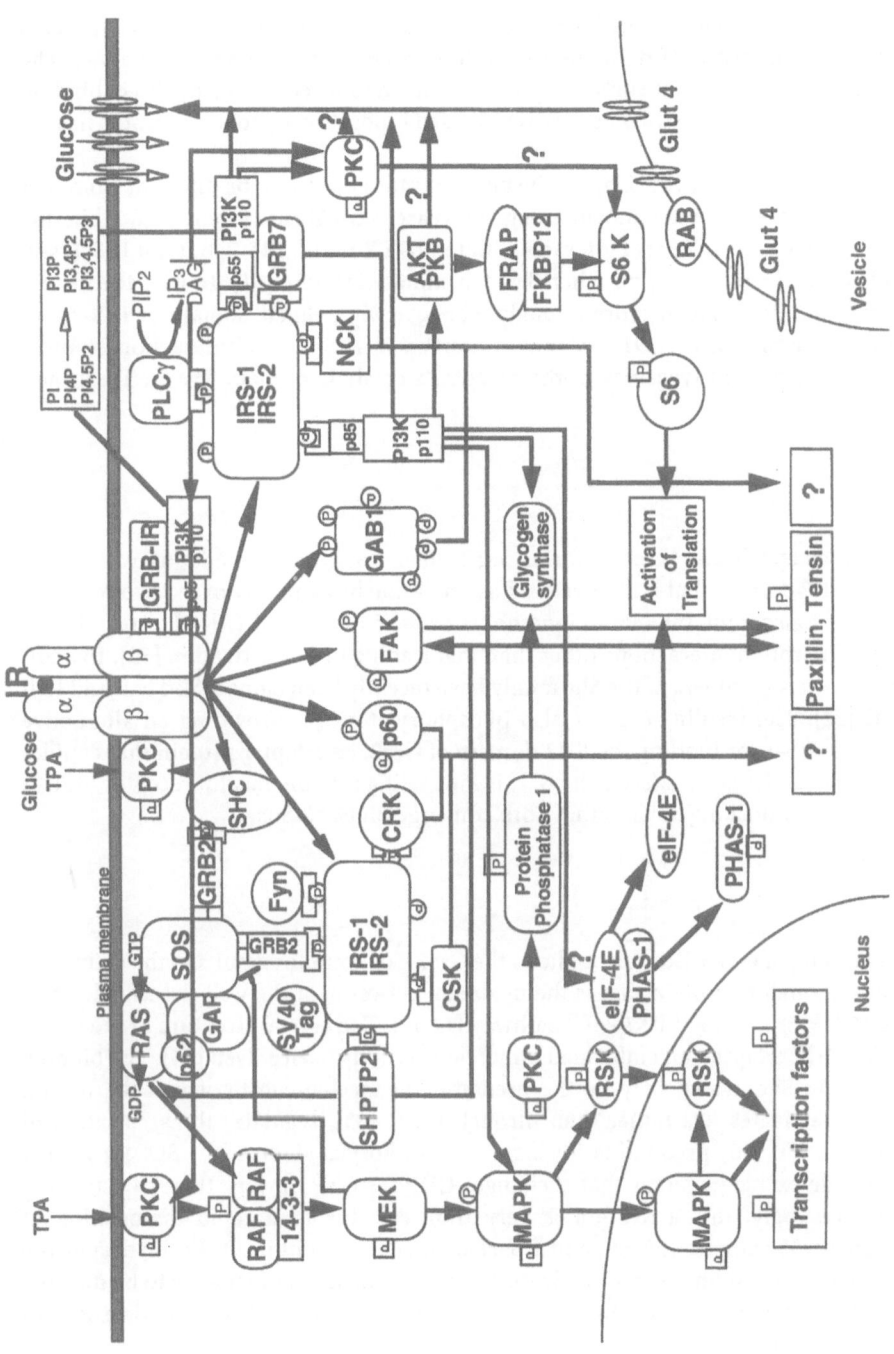

Fig. 2. The insulin signalling pathway. (See text and references therein for explanation)

juxtamembrane region of the insulin receptor has recently been solved [49]. It shares a common fold with PH domains and with the PTB domain of Shc (see below). The binding of IRS-1 to the insulin receptor was shown to be enhanced by direct binding of protein-tyrosine phosphatase 1D, which itself binds directly to the activated insulin receptor [50].

The recognition of IRS-2 by the insulin receptor appears to be different from that of IRS-1. It was recently shown using the yeast two-hybrid system that, while the $IH2^{PTB}$ domain of IRS-2 also interacts with the NPEY motif, IRS-2, but not IRS-1, still binds to the insulin receptor when Tyr_{960} is mutated to Phe [51]. In fact, the insulin receptor interacts much more strongly with a newly defined domain of IRS-2 that comprises amino acids 591-586 (not conserved in IRS-1); the interaction involves the phosphorylated tyrosines of the receptor's catalytic domain Tyr_{1146}, Tyr_{1150} and Tyr_{1151} [51].

Shc

Shc, which consists of three isoforms of 52, 46 and 66 kDa, is an adapter with a PTB and an SH2 domain that is involved in Ras activation by a number of receptor tyrosine kinases; it also contains two unique domains called CH1 and CH2 that may be involved in protein interactions other than that leading to Ras activation [52]. In addition, two new members of the Shc family have recently been cloned, Sli/ShcB and Rai/ShcC [52]. The insulin receptor also phosphorylates Shc, providing an alternative target to IRS-1 for binding the SH2 domain of Grb2, an adapter protein that couples to Sos and activates Ras (see below). It is interesting that the insulin receptor utilizes two different docking devices that both converge on Grb2-Sos.

Grb2/Sos

The binding of either IRS-1/2 or Shc to the Grb2-Sos complex results in the activation of Ras, a small G-protein that is the upstream activator of the well-described Raf-1/MAPKK-MEK/MAPK-ERK/p90^{rsk} pathway [12,53]. The crystal structure of the mammalian Grb2 adapter, an embedded dimer, has recently been solved [54]. The binding of Grb2 to Shc, which may be the predominant mechanism by which the insulin receptor activates Ras rather than through IRS-1 [55], regulates the association of Grb2 with SOS [56], predominantly in its unphosphorylated form [57]. SOS is a guanyl nucleotide exchange factor that exchanges GDP for GTP on Ras, thereby activating it. Interestingly, this activation is very transient due to a rapid dissociation of the Grb2-SOS complex [58–60] and is back to basal levels within 20–30 min despite the persistence of insulin receptor activation. This desensitization appears to be due to a feedback mechanism whereby MEK phosphorylates SOS and thereby desensitizes Ras-GTP loading [59].

Raf-1

The mechanism of Raf-1 activation by Ras is complex and involves binding of Raf-1 to Ras in its active GTP-bound form and Raf-1 is thereby recruited to the membrane and is phosphorylated [61]. Recent evidence suggests that oligomerization of Raf through its binding to the dimeric 14-3-3 proteins also contributes to Raf activation [62–64]. The crystal structure of the Ras homologue Rap1A with the Ras-binding domain of Raf-1 (Raf-RBD) has recently been solved [65], and the functional epitope of Raf-1 investigated by mutagenesis based on that structure [66].

Other Proteins Associated with IRS 1/2

Besides Grb2, a number of other SH2-containing proteins associate with IRS-1/2: the p85α/β subunit isoforms of phosphatidylinositol-3-kinase (PI3)-kinase, the phosphatase SH-PTP2 or Syp, and the adapters Nck [32] and Crk (M.G. Myers, personal communication). IRS-1 has also been reported to associate with the SV40 large T antigen [67], with 14-3-3 proteins [10], with β3-integrins [10], with the tyrosine kinase p59 fyn [68], with a new member of the peptidyl-prolyl cis/trans isomerase (PPIase) family [69], and with the SERCA1 Ca^{2+}-ATPase from rat skeletal muscle [70]. IRS-2 has been shown to interact with Tyk-2 [71] during IFN-α stimulation.

p85/PI3-Kinase

The p85 appears to function as an adapter protein that mediates the association to IRS-1 of not only the 110 kDa catalytic subunit of PI3-kinase, but also GAP and its p62 associated protein [72]. It is also reported to bind and be activated by c-fyn [73] and Ras [74]. The crystal structure of p85 amino-terminal SH2 domain with its phosphopeptide complexes has recently been reported [75]. A new molecule related to the α/β isoforms of the p85 subunit of PI3-kinase, p55^{PIK} has recently been cloned [76]. It appears to be expressed and required during neuronal differentiation [77]. Recent work has provided strong evidence for a role for PI3-kinase in activating p70^{s6k} (reviewed in [78]). Protein kinase B (PKB, the cellular homologue of the transforming kinase v-Akt, or Rac-PK), was shown to be activated by insulin [79] and to lie downstream of PI3-kinase but upstream of p70^{S6K} [80]. PI3-phosphate may bind to the PH domain of PKB [81]. A constitutively activated PI3-kinase has been shown to activate Ras in *Xenopus* oocytes [82].

SH-PTP2/Syp

The phosphatase SH-PTP2 or Syp has been shown to be required as a positive effector for insulin signalling through the MAP kinase pathway [83,84], probably by acting downstream of Ras [85], although maybe not through its binding to IRS-1 but rather through direct binding to the insulin receptor [32,86]. The crystal structure of the two tandem SH2 domains of SHPTP2, which are simultaneously occupied by two IRS-1 phosphorylation sites during stimulation [87], was solved [88].

Other Downstream Signalling Molecules

The adapter Nck, the function of which was not established, was recently found to bind to SOS and to activate Ras-dependent gene expression [89]. Among the more recently identified downstream targets of insulin action, we should mention PHAS-I and II, members of a newly discovered family of proteins that bind eIF-4E and inhibit translation initiation [90,91]; focal adhesion kinase, which is unique in that it is dephosphorylated (with paxillin as a substrate and an interaction with Crk's substrate p130^{CAS}) [87,92–94]; annexin II [95]; C-cbl [96]; caveolin [97] and a new p58/p53 protein with weak homology to two yeast proteins, BOB1 and BEB1 [98].

As is clear from the above description, significant progress has recently been achieved in the dissection of the insulin signalling pathways, including the resolution of the three-dimensional structure of several important recognition domains of signalling molecules. From a physiological viewpoint, however, some questions have arisen regarding the relevance of what has been generally accepted as the two major axes of the insulin signalling cascade, the Ras-MAP kinase cascade and the PI3-kinase pathway.

Are MAP Kinase and PI3-Kinase Essential for Insulin Action?

The Ras-MAP kinase pathway, which has been amazingly conserved in yeast, worm and vertebrates, has been hailed as an "apparently universal signalling pathway through which many of the signals that direct nucleated cells to divide and differentiate are channelled", and the resulting "satisfaction of reducing extraordinarily complex phenomenology down to simple, apparently universally relevant truths" has been stressed [99].

As far as insulin metabolic actions are concerned, however, the role of the MAP kinase pathway has been seriously challenged by recent evidence. Insulin clearly induces activation of MAP kinase through various pathways that involve Grb2, SH-PTP2, PI3-kinase and PLC-γ [100]. MAP kinase was previously defined as the main pathway leading through p70^{S6K} to the activation of glycogen synthase [101]. The use of a specific inhibitor of MEK, PD098059 [102] has, however, failed to inhibit this effect of insulin [103]. Moreover, the overexpression of a mutant SOS lacking the guanine-nucleotide exchange domain in CHO-IR cells markedly impaired insulin-stimulated formation of GTP-bound Ras and activation of MAP kinase without affecting activation of glycogen synthase and p70^{s6k} by insulin [104]. The overexpression of a catalytically inactive mutant of SH-PTP2 (with Cys 459 mutated to Ser) abolished the effect of insulin on Ras and MAP kinase but failed to alter the stimulation by insulin of glucose uptake [104]. The introduction of other dominant negative molecules into 3T3L1 adipocytes has also shown that MAP kinase is not involved in insulin-induced glucose uptake [105–107].

Another putative target of the MAP kinase pathway is the regulation of protein synthesis by insulin. A limiting factor in the initiation phase of translation is the mRNA CAP binding protein eIF-4E. PHAS-I (and -II) bind eIF-4E and inhibit translation initiation. PHAS-I is an important mediator of insulin effects on protein synthesis and was initially thought to be a link between MAP kinase and translation initiation [90]. PHAS-l is indeed a good substrate of MAP kinase in vitro, but several recent lines of evidence indicate that MAP kinase does not mediate the phosphorylation of PHAS-I in response to insulin in cells, but rather that insulin and growth factors increase the phosphorylation of PHAS-I via a rapamycin-sensitive pathway that involves the target of rapamycin, mTOR (also named RAFT or FRAP) [90,103].

Other data have recently also suggested that MAP kinase may not be involved in the mitogenic effects of insulin in a T-lymphoma cell line devoid of IGF-I receptors [40]. Other insulin effects in which MEK inhibition failed to suppress insulin action include lipogenesis and amino acid uptake in rat adipocytes, and the effects of insulin on phosphoenolpyruvate-carboxykinase (PEPCK) and glucokinase gene expression in hepatocytes (A.R. Saltiel, personal communication). In contrast, PD098059 did inhibit insulin effects on c-fos induction, p90^{rsk}, DNA synthesis, and ras desensitization (A.R. Saltiel, personal communication). The role of MAP kinase in insulin action has also been critically reviewed in [108].

The role of PI3-kinase on glucose transport, and more specifically on GLUT4 translocation [109,110], is supported by several lines of evidence. Wortmannin and LY90024, two inhibitors of PI3-kinase, inhibited glucose transport in fat cells [111–113]. Microinjection of antiphosphotyrosine antibodies or inhibition of endogenous PI3-kinase by microinjection of a GST-p85 SH2 fusion protein markedly inhibited insulin-stimulated GLUT4 translocation [105]. Overexpression of a mutant p85 that lacks a binding site for p110 in CHO-IR cells markedly impaired insulin-stimulated accumulation of PI3,4,5-P$_3$ and the insulin-stimulated glucose uptake due to the translocation of GLUT1 [104]. Some controversy arose when three groups reported apparently discrepant results regarding the effect of PDGF, a potent stimulator of PI3-kinase, on GLUT4 translocation [114–116] (reviewed in [11]).

The key to the controversy may have to do with the compartmentalization of cellular signals, a topic that may be crucial for the understanding of signalling specificity (see below). Brozinick et al. have shown recently that overexpressed GLUT4 does not participate proportionally in GLUT4 subcellular trafficking in skeletal muscle of transgenic mice [117]. Moreover, studies in skeletal muscle have identified insulin-sensitive as well as insulin-insensitive intracellular GLUT4-containing membranes [118].

In agreement with the hypothesis that individual growth factors may differentially regulate PI3-kinase activity at specific local sites and thus allow them to mediate different cellular responses, two groups found recently that insulin is much more effective than PDGF in stimulating PI3-kinase activity in the microsomal fraction where the pool of translocatable GLUT4 transporters resides [119,120].

While the consensus therefore appears to be that PI3-kinase is required for insulin-induced glucose transport, it remains to be demonstrated that the kinase activation is sufficient.

Signalling Crosstalk and the Paradox of Signalling Specificity

As initial studies suggested that IRS-1 was specifically engaged by insulin and IGF-I but not by other growth factors such as EGF and PDGF, it was hoped that it would provide a selective mechanism for explaining the specificity of insulin's metabolic effects [32]. It soon became clear, however, that IRS 1/2 are also engaged by an increasing number of cytokine receptors, i.e. those for growth hormone [121–123], interleukins 2, 4, 7, 9, 13 and 15, leukaemia inhibitory factor (LIF), and type I interferons (IFNα, β and ω) [32,71,124–126]. Conversely, insulin was recently reported to induce the phosphorylation of JAK1 (thus far considered rather specific for cytokine signalling) through JAK1's association with Grb2 [127].

More recently, G protein-coupled receptors have also entered the picture. In heart muscle of intact animals, angiotensin II was found to stimulate both IRS-1 and IRS-2 phosphorylation to a level of 30%–50% of that achieved by insulin; this effect appears to be mediated via JAK2 kinase which can co-precipitate with IRS-1 and the angiotensin II type 1 receptor ([128], C.R. Kahn, personal communication). In vascular smooth muscle cells there is no stimulation of IRS-1 phosphorylation by angiotensin II, but, in fact, an inhibition of insulin-stimulated phosphorylation. In both heart and vascular smooth muscle cells angiotensin II inhibited insulin-stimulated PI3-kinase activation ([129], C.R. Kahn, personal communication).

The crosstalk between G-coupled receptors and tyrosine kinase receptor pathways through G-protein $\beta\gamma$ subunits is another exciting recent development. $\beta\gamma$ appears to activate both the MAP kinase and PI3-kinase pathways, through Shc, Ras, Grb2-SOS and PKC-dependent pathways (this one via α subunits), presumably through $\beta\gamma$ binding to PH domains of signalling proteins [130–145].

The extensive crosstalk between the signalling pathways of various classes of receptors makes it harder and harder to grasp how signalling pathways maintain some kind of selectivity. This paradox has been discussed elsewhere [8,9] and here we will only briefly stress a couple of relevant issues.

First, the molecular nature of signalling proteins may not be sufficient to determine signalling specificity, due to the relative molecular "promiscuity" of the interfaces involved. The kinetic aspects (the "time factor" [8]) of their activation and inactivation need to be considered [28]. In other words, the descriptive anatomy of intracellular signalling which has made staggering progress in recent years must now be followed up by a quantal leap in understanding its cellular physiology.

Second, the intracellular compartmentalization and targeting of signalling molecules (as discussed above in relation to PI3-kinase and GLUT4 translocation) are also being newly appreciated as essential determinants of signalling specificity [146–148]; they need to be carefully dissected, using physiologically relevant targets. Anchoring proteins may play an essential role in this respect [149].

Targeted Disruption of the Insulin Receptor Gene and Other Diabetogenes

The successful knock-out of several important genes in the insulin signal transduction pathway has recently been accomplished, in each case with some surprises in the phenotypes obtained (see also review in [150]). The first was the targeted disruption of the IRS-1 gene, accomplished by two groups [35,37,151,152]. The mice that were homozygous for disruption of the gene showed only mild insulin resistance, a finding that was unexpected considering the postulated central role of IRS-1 in insulin signalling and that led to the discovery of IRS-2 and the redundancy (or complementarity) between their actions [34]. The most striking feature in the mice was their embryonal and postnatal growth retardation, stressing not only the importance of IRS-1 for IGF's actions but also the possible role of the insulin receptor itself in embryonic growth [153].

Recently, two groups also successfully knocked out the insulin receptor gene [154,155]. In this case, the mice were born with normal size, but died within a few days from diabetic ketoacidosis; they also showed marked postnatal growth retardation and other metabolic disorders including hyperinsulinaemia. This phenotype would not be surprising a priori, considering the primary role of the insulin receptor in insulin action, if it were not for the fact that some human patients had been reported who were homozygous for a null mutation in the insulin receptor gene, and these presented with a leprechaun phenotype (for review, see [154], [155]) and fared far better than the mice postnatally. Interestingly, the heterozygous mice had normal glucose tolerance. The mice carrying the insulin receptor gene knockout had normal birth weight, probably due to upregulation of the IGF-I receptor, but embryos nulligous for both the IGF-I receptor and the insulin receptor genes were only 30% of normal size, half the size of the animals with only IGF-I receptor knock-out [153], again stressing the role of the insulin receptor in embryonic development. A third important gene knockout in the last year was that of GLUT4 [156,157]. Surprizingly, the knockout mice exhibited only small changes in their steady-state blood glucose levels and their ability to dispose of an oral glucose load; they were thus not diabetic although they appeared, like the IRS-1 knockout mice, to be moderately insulin resistant. They were, however, growth retarded and exhibited decreased longevity associated with cardiac hypertrophy and severely reduced adipose tissue deposits [156].

These results support the concept that glucose homeostasis is a sturdy and complex process, with built-in redundancies and complementarities in the pathways involved. They also support the "diabetogenes concept" of NIDDM whereby combinations of several gene defects in the main metabolic pathways that regulate β-cell secretion and determine peripheral insulin sensitivity are required to reach a threshold above which glucose homeostasis collapses [158]. This concept has recently been given strong experimental support by successful attempts to create new animal models of NIDDM by engineering mice combining multiple heterozygous gene disruptions in insulin signalling pathways and β-cell secretory pathways. Bruening et al. [159] bred heterozygotes of the insulin receptor and IRS-1 knockout mice discussed above. They obtained double heterozygous animals which exhibited normal intrauterine and postnatal development. Although all groups were normoglycaemic, insulin levels were elevated fivefold in IR (+/−) mice and 13-fold in the double heterozygotes, indicating a cumulative effect of the gene defects on insulin resistance. Furthermore, there was a cumulative β-cell hyperplasia, with a three- to fourfold increase in β-cell mass in single heterozygotes and sevenfold in the double heterozygote, the latter showing distorted islet morphology [159].

Terauchi et al. [160] managed to create a more overt NIDDM phenotype with a different combination of heterozygous mutations. They made double heterozygotes for IRS-1 and β-cell glucokinase null mutations. Homozygous IRS-1 gene knockouts showed normal glucose tolerance, fasting hyperinsulinaemia and selective β-cell hyperplasia. Heterozygous mice with β-cell glucokinase gene knockout alone show impaired glucose tolerance due to decreased insulin secretion to glucose. The double knockout mice (IRS-1/glucokinase) developed overt diabetes. They showed fasting hyperinsulinaemia and selective β-cell hyperplasia like the IRS-1 knockouts, but impaired insulin secretion in response to glucose. Thus, the coexistence of two single mutations that were non-diabetogenic by themselves caused overt diabetes, providing the first reconstitution of NIDDM as a polygenic disorder in mice.

Conclusions

Truly spectacular progress has been accomplished over the last few years in our understanding of the downstream signalling pathways mediating insulin actions. As the network unravels, however, it appears that our comprehension of signalling specificity becomes more elusive since the pathways being defined are shared by multiple extracellular signals and, moreover, show extensive crosstalk. It is becoming clear that to shed light into the real physiology of signalling, new dimensions of the signalling network will have to be carefully explored, such as the kinetics of activation of signalling elements, feedback mechanisms and cellular compartmentalization of signals.

Acknowledgements. Christine Reynet, Emmanuel Van Obberghen, Martin Myers and C. Ronald Kahn are gratefully acknowledged for clarifying discussions, and Steve Shoelson, Alan R. Saltiel and Emmanuel Van Obberghen for providing preprints and data before publication. The Hagedorn Research Institute is a basic research component of Novo Nordisk.

References

1. Hubbard SR, Wei L, Ellis L, Hendrickson WA (1994) Crystal structure of the tyrosine kinase domain of the human insulin receptor. Nature 372: 746–754
2. McDonald NQ, Murray-Rust J, Blundell TL (1995) The first structure of a receptor tyrosine kinase domain: a further step in understanding the molecular basis of insulin action. Structure 3: 1–6
3. Taylor SS, Radzio-Andzelm E, Hunter T (1995) How do protein kinases discriminate between serine/threonine and tyrosine? Structural insights from the insulin receptor protein-tyrosine kinase. FASEB J 9: 1255–1266
4. Lee J, Pilch PF (1994) The insulin receptor: structure, function and signalling. Am J Physiol 266: C319–C334
5. De Meyts P (1994) The structural basis of insulin and insulin-like growth factor-I (IGF-I) receptor binding and negative cooperativity, and its relevance to mitogenic versus metabolic signalling. Diabetologia 37 [Suppl 2]: S135–S148
6. Schäffer L (1994) A model for insulin binding to the insulin receptor. Eur J Biochem 221: 1127–1132
7. De Meyts P, Wallach B, Christoffersen CT, Ursø B, Grønskov K, Latus LJ, Yakushiji F, Ilondo MM, Shymko RM (1994) The insulin-like growth factor-I receptor. Structure, ligand binding mechanism and signal transduction. Horm Res 42: 152–169
8. De Meyts P, Christoffersen CT, Ursø B, Wallach B, Grønskov K, Yakushiji F, Shymko RM (1995) Role of the time factor in signalling specificity. Application to mitogenic and metabolic signalling by the insulin and insulin-like growth factor-I receptor tyrosine kinases. Metabolism 44 [Suppl 4]: 1–11
9. De Meyts P, Ursø B, Christoffersen CT, Shymko RM (1995) Mechanism of insulin and IGF-I receptor activation and signal transduction specificity. Receptor dimer cross-linking, bell-shaped curves, and sustained versus transient signalling. Ann N Y Acad Sci 766: 388–401
10. Myers MG, White MF (1996) Insulin signal transduction and the IRS proteins. Annu Rev Pharmacol Toxicol 36: 615–658
11. De Meyts P, Christoffersen CT, Tornqvist H, Seedorf K (1996) Insulin receptors and insulin action. Curr Opin Endo Metab 3: 369–377

12. Seedorf K (1995) Intracellular signalling by growth factors. Metabolism 44 [Suppl 4]: 24–32
13. Schlessinger J (1988) Signal transduction by allosteric receptor oligomerization. Trends Biochem Sci 13: 443–447
14. Yarden Y, Ullrich A (1988) Growth factor receptor tyrosine kinases. Annu Rev Biochem 57: 443–478
15. Metzger H (1992) Transmembrane signalling: the joy of aggregation. J Immunol 149: 1477–1487
16. Heldin C-H, Östman A (1996) Ligand-induced dimerization of growth factor receptors: variations on the theme. Cytokine Growth Factor Rev 7: 3–10
17. Wells JA, de Vos AM (1996) Hematopoietic receptor complexes. Annu Rev Biochem 65: 609–634
18. de Vos AM, Ultsch M, Kossiakoff AA (1992) Human growth hormone and extracellular domain of its receptor: crystal structure of the complex. Science 255: 306–312
19. Ihle JN (1995) Cytokine receptor signalling. Nature 377: 591–594
20. Zhong P, Cara JF, Tager HS (1993) Importance of receptor occupancy, concentration differences, and ligand exchange in the insulin-like growth factor-I receptor system. Proc Natl Acad Sci USA 90: 11451–11455
21. Christoffersen CT, Bornfeldt KE, Rotella CM, Gonzales N, Vissing H, Shymko RM, ten Hoeve J, Groffen J, Heisterkamp N, De Meyts P (1994) Negative cooperativity in the insulin-like growth factor-I (IGF-I) receptor and a chimeric IGF-I/insulin receptor. Endocrinology 135: 472–475
23. Fuh G, Cunningham BC, Fukunaga R, Nagata S, Goeddel DV, Wells JA (1992) Rational design of potent antagonists to the growth hormone receptor. Science 256: 1677–1680
24. De Meyts P, Roth J, Neville Jr. DM, Gavin III JR, Lesniak MA (1973) Insulin interactions with its receptors: experimental evidence for negative cooperativity. Biochem Biophys Res Commun 55: 154–161
25. De Meyts P, Bianco AR, Roth J (1976) Site-site interactions among insulin receptors: characterization of the negative cooperativity. J Biol Chem 251: 1877–1888
26. De Meyts P, Van Obberghen E, Roth J, Brandenburg D, Wollmer A (1978) Mapping of the residues of the receptor binding region of insulin responsible for the negative cooperativity. Nature 273: 504–509
27. De Meyts P, Christoffersen CT, Ursø B, Ish-Shalom D, Sacerdotti-Sierra N, Drejer K, Schäffer L, Shymko RM, Naor D (1993) Insulin potency as a mitogen is determined by the half-life of the insulin-receptor complex. Exp Clin Endocrinol 101: 22–23
28. Hansen BF, Danielsen GM, Drejer K, Sorensen AR, Wiberg FC, Klein HH, Lundemose AG (1996) Sustained signalling from the insulin receptor after stimulation with insulin analogues exhibiting increased mitogenic potency. Biochem J 315: 271–279
29. Hebert TE, Moffett S, Morello J-P, Loisel TP, Bichet DG, Barret C, Bouvier M (1996) A peptide derived from a beta2-adrenergic receptor transmembrane domain inhibits both receptor dimerization and activation. J Biol Chem 271: 16384–16392
30. Limbird L, De Meyts P, Lefkowitz RJ (1975) Beta-adrenergic receptors: evidence for negative cooperativity. Biochem Biophys Res Commun 64: 1160–1168
31. De Meyts P (1976) Cooperative properties of hormone receptors in cell membranes. J Supramol Struct 4: 241–258
32. Myers MG, White MF (1995) New frontiers in insulin receptor substrate signalling. Trends Endocrinol Metab 6: 209–215
33. White MF (1996) The IRS signalling system in insulin and cytokine action. Philos Trans R Soc Lond B Biol Sci 351: 181–189
34. Sun XJ, Wang LM, Zhang Y, Yenush L, Myers MGJ, Glasheen E, Lane WS, Pierce JH, White MF (1995) Role of IRS-2 in insulin and cytokine signalling. Nature 377: 173–177
35. Araki E, Lipes M, Patti ME, Brüning JC, Haag III B, Johnson RS, Kahn CR (1994) Alternative pathway of insulin signalling in mice with targeted disruption of the IRS-1 gene. Nature 372: 186–190
36. Patti ME, Sun XJ, Bruening JC, Araki E, Lipes MA, White MF, Kahn CR (1995) 4PS/insulin receptor substrate (IRS)-2 is the alternative substrate of the insulin receptor in IRS-1 deficient mice. J Biol Chem 270: 24670–24673
37. Tobe K, Tamemoto H, Yamauchi T, Aizawa S, Yazaki Y, Kadowaki T (1995) Identification of a novel substrate for the insulin receptor kinase functionally similar to insulin receptor substrate-1. J Biol Chem 270: 5698–5701

38. Wang L, Hayashi H, Mitani Y, Ishii K, Ohnishi T, Niwa Y, Kido H, Ebina Y (1995) Cloning of a cDNA encoding a 190-kDa insulin receptor substrate-1-like protein of simian COS cells. Biochem Biophys Res Commun 216: 321–328

39. Holgado-Madruga M, Emlet DR, Moscatello DK, Godwin AK, Wong AJ (1996) A Grb2-associated docking protein in EGF- and insulin receptor signalling. Nature 379: 560–564

40. Ursø B, Christoffersen CT, Ouwens M, Vlahos C, Tornqvist C, White MF, Naor D, De Meyts P (1996) Mitogenic signalling by insulin in a T-cell lymphoma, the LB cell line, devoid of IGF-1 receptors: evidence for the lack of involvement of the Ras-MAP kinase pathway and for a possibly novel IRS-like molecule. Exp Clin Endocrinol Diabetes 104: 52–53

41. Lemmon MA, Ferguson KM, Schlessinger J (1996) PH domains: diverse sequences with a common fold recruit signalling molecules to the cell surface. Cell 85: 621–624

42. Shaw G (1996) The pleckstrin homology domain – an intriguing multifunctional protein module. BioEssays 18: 35–46

43. Ferguson KM, Lemmon M, Schlessinger J, Sigler P (1994) Crystal structure at 2.2 Å resolution of the pleckstrin homology domain from human dynamin. Cell 79: 199–209

44. Ferguson KM, Lemmon M, Schlessinger J, Sigler PB (1995) Structure of the high affinity complex of inositol triphosphate with a PLC pleckstrin homology domain. Cell 83: 1037–1046

45. Hyvonen M, Macias MJ, Nilges M, Oshkinat H, Saraste M, Wilmanns M (1995) Structure of the binding sites for inositol phosphates in a PH domain. EMBO J 14: 4676–4685

46. Myers MGJ, Grammer TC, Brooks J, Glasheen EM, Wang LM, Sun X, Blenis J, Pierce J, White MF (1995) The pleckstrin homology domain in IRS-1 sensitizes insulin signalling. J Biol Chem 270: 11715–11718

47. Voliovitch H, Schindler DG, Hadari YR, Taylor SI, Zick Y (1995) Tyrosine phosphorylation of insulin receptor substrate-1 in vivo depends upon the presence of its pleckstrin homology region. J Biol Chem 270: 18083–18087

48. Conway BR, Heller-Harrison RA, Sleeman MW, Czech MP (1996) The pleckstrin homology (PH) domain of IRS-1 is required for tyrosine phosphorylation of IRS-1 but is not essential for membrane localization or insulin-dependent release into cytosol. Exp Clin Endocrinol Diabetes 104 [Suppl 2]: 35–36

49. Eck MJ, Dhe-Paganon SD, Trüb T, Nolte R, Shoelson SE (1996) Structure of the IRS-1 PTB domain bound to the juxtamembrane region of the insulin receptor. Cell 85: 695–705

50. Kharitonenkov A, Schnekenburger J, Chen Z, Knyazev P, Ali S, Zwick E, White MF, Ullrich A (1995) Adapter function of protein-tyrosine phosphatase 1D in insulin receptor/insulin receptor substrate-1 interaction. J Biol Chem 270: 29189–29193

51. Sawka-Verhelle D, Tartare-Deckert S, White MF, Van Obberghen E (1996) Insulin receptor substrate-2 binds to the insulin receptor through its phosphotyrosine-binding domain and through a newly identified domain comprizing amino-acids 591–786. J Biol Chem 271: 5980–5983

52. Bonfini L, Migliaccio E, Pelicci G, Lanfrancone L, Pelicci P (1996) Not all Shc's roads lead to Ras. Trends Biochem Sci 21: 257–261

53. Seger R, Krebs EG (1995) The MAPK signalling cascade. FASEB J 9: 726–735

54. Maignan S, Guilloteau J-P, Fromage N, Arnoux B, Becquart J, Ducruix A (1995) Crystal structure of the mammalian Grb2 adaptor. Science 268: 291–293

55. Pruett W, Yuan Y, Rose E, Batzer AG, Harada N, Skolnick EY (1995) Association between Grb2/Sos and insulin receptor substrate 1 is not sufficient for activation of extracellular signal-regulated kinases by interleukin-4: implications for Ras activation by insulin. Mol Cell Biol 15: 1778–1785

56. Ravichandran K, Lorenz U, Shoelson S, Burakoff S (1995) Interaction of Shc with Grb2 regulates association of Grb2 with mSOS. Mol Cell Biol 15: 593–600

57. de Vries-Smits AMM, Pronk GJ, Medema JP, Burgering BMT, Bos JL (1995) Shc associates with an unphosphorylated form of the p21ras guanine nucleotide exchange factor mSOS. Oncogene 10: 919–925

58. Cherniack AD, Klarlund JK, Conway BR, Czech M (1995) Disassembly of Son-of-sevenless proteins from Grb2 during p21 ras desensitization by insulin. J Biol Chem 270: 1485–1488

59. Waters SB, Holt KH, Ross SE, Syu L-J, Guan K-L, Saltiel AR, Koretzky GA, Pessin JE (1995) Desensitization of Ras activation by a feedback disassociation of the SOS-Grb2 complex. J Biol Chem 270: 20883–20886

60. Langlois WJ, Sasaoka T, Saltiel AR, Olefsky JM (1995) Negative feedback regulation and desensitization of insulin-stimulated and epidermal growth factor stimulated p21 ras activation. J Biol Chem 270: 25320–25323

61. Marais R, Light Y, Paterson HF, Marshall CJ (1995) Ras recruits Raf-1 to the plasma membrane for activation by tyrosine phosphorylation. EMBO J 14: 3136–3145

62. Marshall CJ (1996) Raf gets it together. Nature 383: 127–128

63. Luo Z, Tzivion G, Belshaw PJ, Vavvas D, Marshall M, Avruch J (1996) Oligomerization activates c-Raf-1 through a Ras-dependent mechanism. Nature 383: 181–185

64. Farrar MA, Alberola-Ila J, Perlmutter R (1996) Activation of the Raf-1 kinase by coumermycin-induced dimerization. Nature 383: 179–181

65. Nassar N, Horn G, Herrmann C, Scherer A, McCormick F, Wittinghofer A (1995) The 2.2 Å crystal structure of the Ras-binding domain of the serine/threonine kinase c-Raf-1 in complex with Rap1A and a GTP analogue. Nature 375: 554–560

66. Block C, Janknecht R, Hermann C, Nassar N, Wittinghofer A (1996) Quantitative structure-analysis correlating Ras/Raf interaction in vitro to Raf activation in vivo. Nature Struct Biol 3: 244–251

67. Fei ZL, Dambrosio C, Li SW, Surmacz E, Baserga R (1995) Association of insulin-receptor substrate-1 with simian virus large T antigen. Mol Cell Biol 15: 4232–4239

68. Sun XJ, Pons S, Asano T, Myers MG, Glasheen E, White MF (1996) The fyn tyrosine kinase binds IRS-1 and forms a distinct signalling complex during insulin stimulation. J Biol Chem 271: 10583–10587

69. Miyaji I, Niederfellner G, White MF (1996) IRS-1 and IRS-2 interact with a new member of the peptidyl-prolyl cis/trans isomerase (PPIase) family. Diabetes 45 [Suppl 2]: 438

70. Algenstadt P, Antonetti DA, Kahn CR (1996) Novel interactions of IRS-1 in human skeletal muscle. Exp Clin Endocrinol Diabetes 104 [Suppl 2]: 145–146

71. Platanias LC, Uddin S, Yetter A, Sun XJ, White MF (1996) The type 1 interferon receptor mediates tyrosine phosphorylation of insulin receptor substrate-2. J Biol Chem 271: 278–282

72. Sung CK, Sanchez-Margalet V, Goldfine I (1994) Role of p85 subunit of phosphatidylinositol-3-kinase as an adaptor molecule linking the insulin receptor, p62, and GTPase-activating protein. J Biol Chem 269: 12503–12507

73. Pleiman CR, Hertz WM, Cambier JC (1994) Activation of phosphatidylinositol 3′-kinase by Src-family kinase SH3 binding to the p85 subunit. Science 263: 1609–1612

74. Rodriguez-Viciana P, Warne PH, Dhand R, Van Haesenbroeck B, Gout I, Fry MJ, Waterfield MD, Downward J (1994) Phosphatidylinositol-3-OH kinase as a direct target of Ras. Nature 370: 527–532

75. Nolte RT, Eck MJ, Schlessinger J, Shoelson SE, Harrison SC (1996) Crystal structure of the PI-3 kinase p85 amino-terminal SH2 domain and its phosphopeptide complexes. Nature Struct Biol 3: 364–374

76. Pons S, Asano T, Glasheen E, Miralpeix M, Weiland A, Zhang Y, Myers MGJ, Sun X, White MF (1995) The structure and function of p55PIK reveals a new regulatory subunit for the phosphatidylinositol-3 kinase. Mol Cell Biol 15: 4453–4465

77. Pons S, White MF (1996) p55PIK is expressed and required during neuronal differentiation. Exp Clin Endocrinol Diabetes 104 [Suppl 2]: 40–41

78. Proud CG (1996) p70 S6 kinase: an enigma with variations. Trends Biochem Sci 21: 181–185

79. Kohn AD, Kovacina KS, Roth RA (1995) Insulin stimulates the kinase activity of RAC-PK, a pleckstrin homology domain-containing ser/thr kinase. EMBO J 14: 4288–4295

80. Burgering BMT, Coffer PJ (1995) Protein kinase B (c-Akt) in phosphatidylinositol-3-OH kinase signal transduction. Nature 376: 599–602

81. Franke TF, Yang SI, Chan TO, Datta A, Kazlaukas A, Morrison DK, Kaplan DR, Tsichlis PN (1995) The protein kinase encoded by the Akt proto-oncogene is a target of the PDGF-activated phosphatidylinositol 3-kinase. Cell 81: 727–736

82. Hu Q, Klippel A, Muslin AJ, Fanti WJ, Williams LT (1995) Ras-dependent induction of cellular responses by constitutively active phosphatidylinositol-3 kinase. Science 268: 100–102

83. Yamauchi K, Milarski KL, Saltiel AR, Pessin JE (1995) Protein-tyrosine phosphatase SHPTP2 is a required positive effector for insulin downstream signalling. Proc Natl Acad Sci USA 92: 664–668

84. Milarski KL, Saltiel AR (1995) Expression of catalytically inactive Syp phosphatase in 3T3 cells blocks stimulation of mitogen-activated protein kinase by insulin. J Biol Chem 269: 21239–21243

85. Sawada T, Milarski KL, Saltiel AR (1995) Expression of a catalytically inert Syp blocks activation of MAP kinase pathway downstream of p21 ras. Biochem Biophys Res Comm 214: 737–743

86. Staubs PA, Reichart DR, Saltiel AR et al (1994) Localization of the insulin receptor binding sites for the SH2 domain proteins p85, syp, and GAP. J Biol Chem 269: 27186–27192

87. Schaller MD, Parsons JT (1995) pp125 FAK-dependent tyrosine phosphorylation of paxillin creates a high affinity binding site for Crk. Mol Cell Biol 15: 2635–2645

88. Eck MJ, Pluskey S, Trüb T, Harrison SC, Shoelson SE (1996) Spatial constraints on the recognition of phosphoproteins by the tandem SH2 domains of the phosphatase SH-PTP2. Nature 379: 277–280

89. Hu Q, Milfay D, Williams LT (1995) Binding of Nck to SOS and activation of Ras-dependent gene expression. Mol Cell Biol 15: 1169–1174

90. Lin TA, Kong X, Haystead TAJ, Pause A, Belsham G, Sonenberg N, Lawrence JCJ (1994) PHAS-1 as a link between mitogen-activated protein kinase and translation initiation. Nature 266: 653–656

91. Lin T-A, Kong X, Pang S, Lawrence JCJ (1996) Hormonal control of protein synthesis mediated by the translational regulators, PHAS-I and PHAS-II. 10th International Congress of Endocrinology, San Francisco, Abstract S25-3

92. Pillay TS, Sasaoka T, Olefsky J (1995) Insulin stimulates the tyrosine dephosphorylation of pp125 focal adhesion kinase. J Biol Chem 270: 991–994

93. Knight JB, Yamauchi K, Pessin J (1995) Divergent insulin and platelet-derived growth factor regulation of focal adhesion kinase (pp125 Fak) tyrosine phosphorylation, and rearrangement of actin stress fibers. J Biol Chem 270: 10199–10203

94. Polte TR, Hanks SK (1995) Interaction between focal adhesion kinase and Crk-associated tyrosine kinase substrate p130 CAS. Proc Natl Acad Sci USA 92: 10678–10682

95. Biener Y, Feinstein R, Mayak M, Kaburagi Y, Kadowaki T, Zick Y (1996) Annexin II – a novel element in insulin signal transduction: role of annexin II phosphorylation in insulin receptor internalization. Exp Clin Endocrinol Diabetes 104 [Suppl 2]: 57–58

96. Ribon V, Mastick CC, Saltiel AR (1996) C-cbl undergoes tyrosine phosphorylation and interacts with specific SH2 domains in response to insulin. Diabetes 45 [Suppl 2]: 107

97. Mastick CC, Brady MJ, Saltiel AR (1995) Insulin specifically stimulates the tyrosine phosphorylation of caveolin in 3T3-L1 adipocytes. Diabetes 44 [Suppl 1]: A 50

98. Yeh TC, Ogawa W, Danielsen AG, Roth RA (1996) Isolation of a cDNA encoding a previously uncharacterized substrate of the insulin receptor family. 10th International Congress of Endocrinology, San Francisco, Abstract P1-31

99. Egan SE, Weinberg RA (1993) The pathway to signal achievement. Nature 365: 781–783

100. Seedorf K, Juhl L, Hansen L, Nielsen L, Mosthaf L (1996) Insulin-induced activation of MAP kinase involves Grb-2, SHPTP2, PI-3 kinase and PLC gamma. Exp Clin Endocrinol Diabetes 104: 138–139

101. Cohen P, Campbell DG, Dent P, Gomez N, Lavoinne A, Nakielny S, Stokoe D, Sutherland C, Traverse S (1992) Dissection of the protein kinase cascades involved in insulin and nerve growth factor action. Biochem Soc Trans 20: 671–674

102. Alessi DR, Cuenda A, Cohen P, Dudley DT, Saltiel AR (1995) PD-098059 is a specific inhibitor of the activation of mitogen-activated protein kinase kinase in vitro and in vivo. J Biol Chem 270: 27489–27494

103. Azpiazu I, Saltiel AR, De Paoli- Roach AA, Lawrence JC (1996) Regulation of both glycogen synthase and PHAS-1 by insulin in rat skeletal muscle involves mitogen-activated protein kinase independent and rapamycin-sensitive pathways. J Biol Chem 271: 5033–5039

104. Kasuga M, Yonezawa K, Hara K, Sakaue M, Noguchi T, Matozaki T (1996) Role of IRS-1 associated proteins in insulin signalling. 10th International Congress of Endocrinology, San Francisco, Abstract S43-1

105. Haruta T, Morris A, Rose D, Nelson G, Mueckler M, Olefsky J (1995) Insulin-stimulated GLUT4 translocation is mediated by a divergent intracellular signalling pathway. J Biol Chem 270: 27991–27994

106. Ogawa W, Sakaue H, Takada M, Kuroda S, Kotani K, Ueno H, Kasuga M (1996) Introduction of dominant negative molecules into 3T3L1 adipocytes using adenovirus vector. Exp Clin Endocrinol Diabetes 104 [Suppl 2]: 132

107. Gnudi L, Frevert EU, Houseknecht KL, Kahn BB (1996) Use of recombinant adenovirus-mediated gene transfer to investigate the effects of dominant negative ras on the activation of PI3 kinase and glucose transport in 3T3L1 adipocytes. Exp Clin Endocrinol Diabetes 104 [Suppl 2]: 132–133

108. Denton RM, Tavaré JM (1995) Does mitogen-activated-protein kinase have a role in insulin action? Eur J Biochem 227: 597–611

109. Stephens JM, Pilch PF (1995) The metabolic regulation and vesicular transport of GLUT4, the major insulin-responsive glucose transporter. Endocrine Rev 16: 529–546

110. Birnbaum MJ, James D (1995) The insulin-regulatable glucose transporter GLUT-4. Curr Opin Endocrinol Diabetes 2: 383–391

111. Ridderstråle M, Tornqvist H (1994) PI 3-kinase inhibitor wortmannin blocks the insulin-like effect of growth hormone in isolated rat adipocytes. Biochem Biophys Res Commun 203: 306–310

112. Okada T, Kawano Y, Sakibara T, Hazeki O, Ui M (1994) Essential role of phosphatidylinositol 3-kinase in insulin-induced glucose transport and antilipolysis in rat adipocytes: studies with a selective inhibitor wortmannin. J Biol Chem 269: 3568–3573

113. Cheatham B, Vlahos CJ, Cheatham L, Wang L, Blenis J, Kahn CR (1994) Phosphatidylinositol 3-kinase activation is required for insulin stimulation of pp70 S6 kinase, DNA synthesis and glucose transporter translocation. Mol Cell Biol 14: 4902–4911

114. Kamohara S, Hayashi H, Todaka M, Kanai F, Ishii K, Imanaka T, Escobedo JA, Williams LT, Ebina Y (1995) Platelet-derived growth factor triggers translocation of the insulin-regulatable glucose transporter (type 4) predominantly through phosphatidylinositol 3-kinase binding sites on the receptor. Proc Natl Acad Sci USA 92: 1077–1081

115. Isakoff SJ, Taha C, Rose E, Marcusohn J, Klip A, Skolnik EY (1995) The inability of phosphatidylinositol 3-kinase activation to stimulate GLUT4 translocation indicates additional signalling pathways are required for insulin-stimulated glucose uptake. Proc Natl Acad Sci USA 92: 10427–10251

116. Wiese RJ, Corley-Mastick C, Lazar DF, Saltiel AR (1995) Activation of mitogen-activated protein kinase and phosphatidylinostol 3′-kinase is not sufficient for the hormonal stimulation of glucose uptake, lipogenesis, or glycogen synthesis in 3T3-L1 adipocytes. J Biol Chem 270: 3442–3446

117. Brozinick JTJ, Reynolds TH, Wilson CM, McCoid SC, Stephenson RW, Gibbs EM, Cushman SM (1996) Overexpressed GLUT4 do not participate proportionally in GLUT4 subcellular trafficking in skeletal muscle of transgenic mice as assessed by bis-mannose photolabelling. Diabetes 45 [Suppl 2]: Abstract 320

118. Zorzano A, Munoz P, Camps M, Mora C, Testar X, Palacin M (1996) Insulin-induced redistribution of GLUT4 glucose carriers in the muscle fiber. Diabetes 45 [Suppl 1]: S70–S81

119. Navé BT, Hayward A, Siddle K, Sheperd PR (1996) Regulation of phosphoinositide 3-kinase by PDGF and insulin in 3T3-L1 adipocytes. Exp Clin Endocrinol Diabetes 104: 55–56

120. Ricort J-M, Tanti J-F, Van Obberghen E, Le Marchand-Brustel Y (1996) Different effects of insulin and platelet-derived-growth-factor on phosphatidylinositol 3-kinase at the subcellular level in 3T3-L1 adipocytes. Eur J Biochem 239: 17–22

121. Souza SC, Frick GP, Yip R, Lobo RB (1994) Growth hormone stimulated tyrosine phosphorylation of insulin receptor substrate-1. J Biol Chem 269: 30085–30088

122. Argetsinger LS, Hsu GW, Myers GW, Billestrup N, White MF, Carter-Su C (1995) Growth hormone, interferon gamma, and leukemia inhibitory factor promoted tyrosyl phosphorylation of insulin receptor substrate 1. J Biol Chem 270: 14685–14692

123. Ridderstråle M, Degerman E, Tornqvist H (1995) Growth hormone stimulates the tyrosine phosphorylation of the insulin receptor substrate-1 and its association with phosphatidylinositol 3-kinase in primary adipocytes. J Biol Chem 270: 3471–3474

124. Johnston JA, Wang L-M, Hanson EP, Sun X-J, White MF, Oakes SA, Pierce JH, O'Shea JJ (1995) Interleukins 2, 4, 7 and 15 stimulate tyrosine phosphorylation of insulin receptor substrates 1 and 2 in cells. J Biol Chem 270: 28527-28530

125. Wang LM, Michieli P, Lie WR, Liu F, Lee CC, Minty A, Sun XJ, Levine A, White MF, Pierce JH (1995) The insulin receptor substrate-1-related 4PS substrate but not the interleukin-2R-gamma chain is involved in interleukin-13-mediated signal transduction. Blood 86: 4218-4227

126. Argetsinger LS, Norstedt G, Billestrup N, White MF, Carter-Su C (1996) Growth hormone, interferon gamma, and leukemia inhibitory factor promote tyrosyl phosphorylation of 4PS/insulin receptor substrate-2. 10th International Congress of Endocrinology, San Francisco, abstract OR7-7

127. Giorgetti-Peraldi S, Peyrade F, Baron V, Van Obberghen E (1995) Involvement of Janus kinases in the insulin signalling pathway. Eur J Biochem 234: 656-660

128. Kahn CR (1996) The ever expanding network of insulin action and its alterations in diabetes. 10th International Congress of Endocrinology, San Francisco, abstract L-3

129. Folli F, Lin Y-W, Kahn CR, Feener EP (1996) Angiotensin II inhibits PI-3 kinase activation induced by multiple tyrosine kinase receptors. 10th International Congress of Endocrinology, San Francisco, abstract P3-143

130. Koch WJ, Hawes BE, Allen LF, Lefkowitz RJ (1994) Direct evidence that Gi-coupled receptor stimulation of mitogen-activated protein kinase is mediated by G beta/gamma activation of p21 ras. Proc Natl Acad Sci 91: 12706-12710

131. van Biesen T, Hawes BE, Koch WJ, Luttrell LM, Lefkowitz RJ (1995) Mitogenic signalling via G-protein-coupled receptors – role for G-beta/gamma and G-alpha subunits of pertussis-toxin sensitive G-proteins. FASEB J 9: A1411

132. Inglese J, Koch WJ, Touhara K, Lefkowitz RJ (1995) G beta/gamma interactions with PH domains and Ras-MAPK signalling pathways. Trends Biochem Sci 20: 151-156

133. Hawes BE, van Biesen T, Koch WJ, Luttrell LM, Lefkowitz RJ (1995) Activation of the Ras/mitogen activated protein kinase pathway by the beta-gamma subunit of G-proteins utilizes a tyrosine kinase and is independent of phosphoinositide hydrolysis. FASEB J 9: A1411

134. Pitcher JA, Touhara K, Payne ES, Lefkowitz RJ (1995) Pleckstrin homology domain-mediated membrane association and activation of the beta adrenergic receptor kinase requires coordinate interaction with G beta/gamma subunits and lipid. J Biol Chem 270: 11707-11710

135. Luttrell LM, Hawes BE, Touhara K, van Biesen T, Koch WJ, Lefkowitz RJ (1995) Effect of cellular expression of pleckstrin homology domains on Gi-coupled receptor signalling. J Biol Chem 270: 12984-12989

136. Touhara K, Koch WJ, Hawes BE, Lefkowitz RJ (1995) Mutational analysis of the pleckstrin homology domain of the beta adrenergic receptor kinase- differential effects on G beta/gamma and phosphatidylinositol 4,5 phosphate binding. J Biol Chem 270: 17000-17005

137. Luttrell LM, van Biesen T, Hawes BE, Koch WJ, Touhara K, Lefkowitz RJ (1995) G beta/gamma subunits mediate mitogen-activated protein kinase activation by the tyrosine kinase insulin-like growth factor-1 receptor. J Biol Chem 270: 16495-16498

138. van Biesen T, Hawes BE, Luttrell DK, Krueger KM, Touhara K, Porfiri E, Sakaue M, Luttrell LM, Lefkowitz RJ (1995) Receptor tyrosine kinase-mediated and G beta/gamma-mediated MAP kinase activation by a common signalling pathway. Nature 376: 781-784

139. Touhara K, Hawes BE, van Biesen T, Lefkowitz RJ (1995) G-protein beta-gamma subunits stimulate phosphorylation of Shc adapter proteins. Proc Natl Acad Sci USA 92: 9284-9287

140. Lefkowitz RJ (1996) G-protein-coupled receptors and receptor kinases – from molecular biology to potential therapeutic applications. Nature Biotechnol 14: 283-286

141. Moxham CM, Malbon CG (1996) Insulin action impaired by deficiency of the G-protein subunit $G_i\alpha_2$. Nature 379: 840-844

142. Rodgers BD, Bernier M, Montrose-Rafizadeh C (1996) Functional insulin receptors upregulate expression of a novel G-protein beta subunit homologue. 10th International Congress of Endocrinology, San Francisco, abstract OR31-5

143. van Biesen T, Hawes BE, Raymond JR, Luttrell LM, Koch WJ, Lefkowitz RJ (1996) Go-protein alpha subunits activate mitogen-activated protein kinase via a novel protein kinase C dependent mechanism. J Biol Chem 271: 1266-1269

144. Luttrell LM, Hawes B, van Biesen T, Luttrell DK, Lefkowitz RJ (1996) An essential role for src tyrosine kinase shc adapter protein complexes in G beta/gamma subunit-mediated mitogen-activated protein kinase activation. J Invest Med 44: A271

145. Hawes BE, Luttrell LM, van Biesen T, Lefkowitz RJ (1996) Phosphatidylinositol 3-kinase is an early intermediate in the G beta/gamma-mediated mitogen-activated protein kinase signalling pathway. J Biol Chem 271: 12133–12136

146. Kublaoui B, Lee J, Pilch PF (1995) Dynamics of signalling during insulin-stimulated endocytosis of its receptor in adipocytes. J Biol Chem 270: 59–65

147. Bevan AP, Drake P, Bergeron JJM, Posner BI (1996) Intracellular signal transduction: the role of endosomes. Trends Endocrinol Metab 7: 13–21

148. Valgeirsdottir S, Eriksson A, Nister M, Heldin CH, Westermark B, Claesson-Welsh L (1995) Compartmentalization of autocrine signal transduction pathways in Sis-transformed NIH 3T3 cells. J Biol Chem 270: 10161–10170

149. Mochly-Rosen D (1995) Localization of protein kinases by anchoring proteins: a theme in signal transduction. Science 268: 247–251

150. Patti ME, Kahn CR (1996) Lessons from transgenic and knockout animals about noninsulin-dependent diabetes mellitus. TEM 7: 311–319

151. Tamemoto H, Kadowaki T, Tobe K, Yagi T, Sakura H, Hayakawa T, Terauchi Y, Ueki K, Kaburagi Y, Satoh S, Sekihara H, Yoshioka S, Horikoshi H, Furata Y, Ikawa Y, Kasuga M, Yasaki Y, Alzawa S (1994) Insulin resistance and growth retardation in mice lacking insulin receptor substrate-1. Nature 372: 182–186

152. Yamauchi T, Tobe K, Tamemoto H, Ueki K, kaburagi Y, Yamamoto-Honda R, Takahashi Y, Yoshizawa F, Aizawa S, Akanuma Y, Sonenberg N, Yazaki Y, Kadowaki T (1996) Insulin signalling and insulin actions in the muscles and livers of insulin-resistant, insulin receptor substrate 1-deficient mice. Mol Cell Biol 16: 3074–3084

153. Efstratiadis A (1996) Genetics of growth: developmental roles of IGF and insulin receptors. Exp Clin Endocrinol Diab 104 [Suppl 2]: 4–6

154. Joshi RL, Lamothe B, Cordonnier N, Mesbah K, Monthioux E, Jami J, Bucchini D (1996) Targeted disruption of the insulin receptor gene in the mouse results in neonatal lethality. EMBO J 15: 1542–1547

155. Accili D, Drago J, Lee EJ, Johnson MD, Cool M, Salvatore P, Asico LD, Jose PA, Taylor SI, Westphal H (1996) Early neonatal death in mice homozygous for a null allele of the insulin receptor gene. Nature Genet 12: 106–109

156. Katz EB, Stenbit AE, Hatton K, DePinho R, Charron MJ (1995) Cardiac and adipose tissue abnormalities but not diabetes in mice deficient in GLUT4. Nature 377: 151–155

157. Mueckler M, Holman G (1995) Homeostasis without a GLUT. Nature 377: 100–101

158. De Meyts P (1993) The diabetogenes concept of NIDDM. Adv Exp Med Biol 334: 89–100

159. Bruening JC, Winnay J, Bonner-Weir S, Accili D, Kahn CR (1996) Creating polygenic models of insulin resistance and diabetes mellitus. J Invest Med 44: A258

160. Terauchi Y, Iwamoto K, Tamemoto H, Komeda K, Ishii C, Kanazawa Y, Asanuma N, Aizawa T, Akanuma Y, Yasuda K, Kodama T, Tobe K, Yazaki Y, Kadowaki T (1997) Development of non-insulin-dependent diabetes mellitus in the double knock-out mice with disuption of insulin receptor substrate-1 and beta-cell glucokinase genes: genetic reconstitution of diabetes as a polygenic disease. J Clin Invest 99:861–866

Cell Biology of Insulin Action on Glucose Transport: Looking Back

S.W. CUSHMAN

The relationship between blood sugar and diabetes mellitus has been known literally for centuries [1]. The preeminent role of insulin per se in the regulation of blood sugar and its role in diabetes was recognized before the actual discovery of insulin itself, but rose to the forefront of medicine with its miraculous reversal of the diabetic state when the first crude preparations were injected into mortally ill patients in the 1920s [2]. Only more recently have the roles of other hormones in overall systemic glucose homeostasis been appreciated, as well as the complex involvement of multiple local and systemic factors, and multiple cells and tissues in the regulation of blood glucose levels and metabolism been identified.

The purpose of this short review is to provide an historical perspective and context for the discovery of the novel mechanism through which insulin is now thought to regulate glucose uptake and metabolism at the glucose transport level, namely, the translocation of glucose transporter proteins from an intracellular storage site to the plasma membrane. Because a large number of excellent reviews have appeared over the years which comprehensively document the biochemistry, cell biology, and molecular biology of glucose transport and transporters, and their regulation, this review will be limited to citation of those papers in the literature which stand out in this author's mind either because of their historical significance or because they illustrate especially interesting aspects of the evolution of the translocation hypothesis which are rarely pointed out in descriptions of the latest scientific developments. While this author's involvement in studies of glucose transport did not begin until several years after a postdoctoral training period in Albert Renold's laboratory under the mentorship of Bernard Jeanrenaud, the perspective taken in the present review is meant to reflect the historical and broad physiological view which Albert believed was so important in biomedical research. Albert would have been delighted to see how the glucose transporter translocation story has ultimately emanated from and simultaneously brought together disciplines ranging from medicine and physiology to biochemistry, cell biology, and molecular biology. It was this author's extraordinary privilege to have been introduced to such a perspective through a personal training experience directly with Albert.

The Glucose Transporter Family

The fundamental role of glucose transport in the complex processes which regulate systemic glucose homeostasis is now well established. Indeed, glucose transport itself is complex in so far as five isoforms of the facilitated diffusion type of transporter specific for glucose (GLUT1, GLUT2, GLUT3, GLUT4, and GLUT7) have been identified [3]. These isoforms are products of separate genes and have specific tissue, cell, and even subcellular distributions. For example, while GLUT1 is almost uniformly distributed among tissues and cells, GLUT2 is found only in cells which function as glucose sensors. These include the β cells of the pancreas, and other cells where glucose equilibration across the cell membrane is essential such as hepatocytes which function in both glucose storage and mobilization depending on the metabolic state of the organism. Hepatocytes have been suggested to contain GLUT7 which is proposed to reside in the endoplasmic reticulum and function in glucose transport from the lumen to the cytosol during glycogenolysis and gluconeogenesis. However, nothing further has been reported beyond its initial description. GLUT3 is the predominant neuronal glucose transporter and the complex nature of the processes which regulate glucose metabolism in the brain is reflected by the presence of GLUT1 in the brain endothelial cells and nonneuronal cells, and GLUT3 in the neurons. Surprisingly, GLUT3 is also found in large quantities in platelets where its function is unknown [4]. GLUT5 was initially described as a facilitated glucose transporter but is now strongly believed to function primarily as a fructose transporter and is found in relatively large amounts in sperm and in the intestine. GLUT6 is a pseudogene containing much of the amino acid sequence of GLUT3.

A structurally homologous but genetically unrelated glucose transporter is that found in tissues where glucose is transported against a concentration gradient: the sodium-dependent glucose transporter [5]. Energy is indirectly required for this glucose transporter to function normally in that a sodium gradient across the plasma membrane must be maintained. This glucose transporter carries out its function in cells in conjunction with facilitated glucose transporters so that the concentration gradient of glucose established by this glucose transporter is dissipated by exit from another part of the cell down a concentration gradient. GLUT2 is frequently the partner in this transcellular glucose transport process although GLUT1 can also function in this manner. The absorbtive cells in the gut and reabsorbtive cells in the renal tubules are obvious sites of this activity.

Glucose Transport and the Plasma Membrane

The focus of this brief historical overview is the GLUT4 isoform which represents the predominant glucose transporter expressed in insulin-responsive tissues and cells, primarily muscle and adipose tissue [3]. This is not to say that GLUT4 is the only insulin-responsive glucose transporter since GLUT1 also responds to insulin in insulin target cells. Even GLUT3 has been observed in the L6 muscle cell line to respond to insulin along with GLUT1 and GLUT4 [6]. These observations imply that cellular machinery is as important as glucose transporter isoform in the glucose transport response to insulin.

Despite the pioneering studies of Banting and Best in the 1920s linking insulin to the regulation of blood glucose levels, only in the 1940s and 1950s did Levine and colleagues propose that the link between insulin and blood glucose might comprise a stimulation by insulin of glucose transport across the plasma membrane of target cells [7]. Crofford and Renold finally provided the first direct evidence in vitro for this fundamental action of insulin in 1965 [8]. Until 1980, considerable attention was focused on establishing the rate-limiting nature of glucose transport for glucose metabolism in adipose tissue and skeletal muscle. The former received particular attention because of the description in 1964 by Rodbell of the elegantly simple collagenase digestion technique for preparing isolated adipose cells [9]. In isolated cell preparations, the extracellular spaces normally separating the plasma membrane of the cell from the substrate delivery compartment are removed. Another elegant experimental technique, the "oil separation technique" for rapidly separating isolated adipose cells from their incubation medium by floatation on dinonylphalate oil, was described by Gliemann and colleagues in 1976 [10]. In combination with the uptake of the nonmetabolizable hexose 3-O-methylglucose, this technique permitted the first quantitative studies of glucose transport activity itself in an insulin target cell and a clear demonstration of insulin's stimulatory effect on the glucose transport V_{max}. Indeed, this author is not aware of even a single report in which an hormonal effect of any kind has been observed on glucose transport K_m.

Reports by Carter and Martin further established that a plasma membrane preparation from insulin-stimulated adipose cells retained an elevated glucose transport rate compared to the same membrane fraction prepared from basal, non-insulin-stimulated cells [11]. Wardzala in our laboratory finally reported the adaptation of a cytochalasin B binding assay for use with plasma membranes prepared from adipose cells and demonstrated directly that insulin increased the number of functional glucose transporters [12]. Cytochalasin B, a fungal metabolite, was originally known for its disruptive effects on microfilaments, eventually shown to occur through its high affinity binding to actin. This substance was almost accidentally found as well to be a potent competitive inhibitor of glucose transport and thus became an important reagent in characterizing the glucose transporter in human erythrocytes where it comprises approximately 5% of the plasma membrane protein [13]. Wardzala's painstakingly achieved adaptation was crucially dependent on the use of an equilibrium binding technique combined with the inclusion of cytochalasin E to competitively inhibit cytochalasin B binding to a second class of high affinity binding sites (presumably actin) also present in the membrane preparations. Even in the insulin-stimulated state, glucose transporters comprise <0.1% of the proteins in the rat adipose cell plasma membrane. The ultimate conclusion of all of this work was that insulin somehow activated glucose transporters in the plasma membrane of insulin target cells, presumably present in some sort of inactive and thus undetectable state in the absence of insulin. Considerable effort went into attempts to recreate this activation phenomenon in isolated plasma membranes prepared from basal cells.

Glucose Transporter Translocation

Early Observations

In 1980, our laboratory at the NIH and Tetsuro Kono's laboratory at Vanderbilt University simultaneously described the initial evidence for the novel mechanism of insulin action on glucose transport now known as the "translocation mechanism" [14,15]. An early schematic representation of this hypothesis is shown in Fig. 1 [16]. In

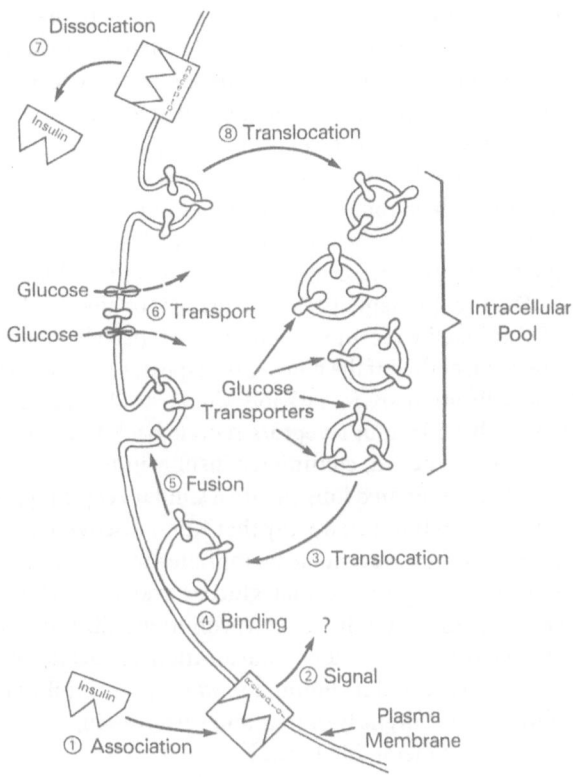

Fig. 1. Hypothetical model of insulin's action on glucose transport published in 1981. (From [16])

the former report, Cushman and Wardzala utilized intracellular membrane and plasma membrane fractions prepared by differential ultracentrifugation and assayed the numbers of glucose transporters by D-glucose-inhibitable cytochalasin B binding; in the latter, Suzuki and Kono used membrane fractions prepared by sucrose density centrifugation and assessed the presence of glucose transporters by a glucose transport reconstitution technique. Perhaps the remarkable coincidence of these two reports using different fractionation techniques and dramatically different glucose transporter assays should have been viewed as a substantially reliable indicator of the ultimate strength of the translocation hypothesis. However, scepticism abounded in part because of the novelty of the putative mechanism itself and in part because of the difficulty of conceiving that such a process could occur fast enough to participate in the rapid glucose uptake response to insulin.

The next several years provided numerous reports describing a wide variety of characteristics of the translocation process; most of these reports were published by our NIH laboratory and Kono's laboratory at Vanderbilt [17]. Translocation was shown to be rapid and reversible, temperature- and energy-dependent, not dependent on protein synthesis, and characterized by an insulin concentration dependency fully consistent with the spare receptor (and signalling) concept also initially demonstrated by Kono and Barham [18]. Insulin target tissues and cells identified with glucose transporter translocation included isolated guinea pig and human adipose cells, and rat diaphragm and heart. Despite an initial report that insulin increased insulin-like growth factor (IGF)-II binding to isolated rat adipose cells by increasing the apparent affinity of the IGF-II receptor for its ligand, this process was also ultimately shown to occur through a translocation process [19]. The physiological significance of this latter translocation process is still unknown; the IGF-II receptor has now been shown to be identical to the mannose-6-phosphate receptor which is thought to function in the transport of lysosomal enzymes from the Golgi apparatus to the lysosomes [20].

During this same period, evidence for the insulin-induced subcellular trafficking of the insulin receptor itself was extensively reported, in this case in the reverse direction of that of the glucose transporter [21]. The characteristics of insulin receptor internalization are remarkably similar to those of glucose transporter exocytosis and internalized insulin receptors recycle back to the plasma membrane following internalization. Indeed, continuous insulin-induced recycling of the insulin receptor was clearly documented long before a similar recycling process was shown for the glucose transporter. It is noteworthy that Kono discovered the translocation of glucose transporters from intracellular membranes to the plasma membrane as a consequence of testing his hypothesis that glucose transporters might actually be internalized in response to insulin in concert with internalization of insulin receptors. Evidence for the regulated subcellular translocation of several other integral membrane proteins was also reported including the water permeability response of toad bladder epithelium to antidiuretic hormone and vasopressin, and acid secretion in gastric mucosa in response to histamine [22,23].

Relationship to Altered Metabolic States

Another series of studies was reported extensively from our laboratory with regard to the mechanisms of the altered glucose transport response in adipose cells from rats in various perturbed metabolic states, especially states of insulin resistance. Thus, Karnieli et al. [24] described a decrease in the intracellular pool of glucose transporters in the adipose cells of streptozotocin-diabetic rats, Hissin et al. [25], a decrease in intracellular glucose transporters in the adipose cells of high fat-fed rats, and Kahn et al. [26], a decrease in the intracellular glucose transporters in adipose cells of fasted rats. In each case, the decrease in intracellular glucose transporters directly corresponded to the decrease observed in the appearance of glucose transporters in the plasma membranes and in the stimulation of glucose transport in the intact cell in response to insulin. Interestingly, the much smaller glucose transport responses to insulin observed in human and guinea pig adipose cells, noted above, correspond to much smaller intracellular pools of glucose transporters in the basal state [27,28]. Similarly, Kahn et al. [29] observed that chronic hyperinsulinemia in normal rats was accompanied by not only an increased glucose transport response to insulin as previously described by others, but also a corresponding increase in the intracellular pool of glucose transporters in the basal state. These observations led to the concept that insulin-resistant and hyper-responsive glucose transport responses in the adipose cell could be explained as simple adaptations to the metabolic state of the animal by decreasing or increasing the number of glucose transporters available in the intracellular pool for recruitment. With the knowledge gained later that GLUT4 is the predominant glucose transporter isoform in the adipose cell, the concept of adaptive down- and up-regulation of the intracellular pool of glucose transporters is not altered by the discovery of multiple glucose transporter isoforms. In addition, this simple concept of an altered intracellular pool of glucose transporters in the basal state circumvents the need for postulating changes in the insulin receptor and its signalling pathways, or in the mechanisms which actually mediate the translocation process.

Obesity and increased adipose cell size, as studied in our laboratory using the aging male Sprague-Dawley rat, presented an unusual circumstance in trying to relate cell geometry, cytoplasmic mass, and the subcellular translocation of glucose transporters [30]. When adipose cells enlarge, the increase in cell surface area expressed as a fold change is less than that of the volume, representing primarily the mass of the large central lipid droplet. Surprisingly, the cytoplasmic mass, as reflected in the intracellular water space, and content of intracellular membranes such as the low-density microsomes containing the intracellular pool of glucose transporters, increase even less as a fold change than the surface area does. Thus the cytoplasm per unit of plasma membrane actually decreases. Total cellular glucose transporter number per cell does not change as cells enlarge but the basal subcellular distribution changes in a way that the concentration of glucose transporters in the plasma membrane stays constant, the total in the plasma membrane increases in proportion to the increase in cell surface area, and the proportion of the total in the intracellular pool decreases. In the largest cells studied, the basal distribution of glucose transporters is actually similar to that in the fully insulin-stimulated state in typical small cells. These largest cells do not respond to insulin.

In contrast, the very large adipose cells present in the young but markedly obese Zucker fatty rat, with a volume approaching that of the largest cells studied in the aging male Sprague-Dawley rat, express a normal amount of cytoplasmic mass per unit of plasma membrane and a markedly increased intracellular pool of glucose transporters compared to cells from young lean littermates [31]. These cells retain a normal basal subcellular distribution of glucose transporters, and normal large glucose transport and glucose transporter translocation responses to insulin. Insulin resistance in the adipose cell with increasing cell size in obesity thus appears to reflect the manner in which the expansion of the intracellular machinery is regulated. The metabolically active adipose cells from young obese Zucker fatty rats retain their insulin responsiveness while the metabolically less active cells from old obese male Sprague-Dawley rats lose their insulin responsiveness in proportion to their intracellular pools of glucose transporters despite their similar cell sizes.

Glucose Transporter Intrinsic Activity

A special problem has accompanied the translocation hypothesis from its very inception: is the intrinsic activity of the glucose transporter also regulated or does it vary? Intrinsic activity is a combined function of glucose transporter turnover number and K_m [10]. Because regulation of transport K_m has never been observed, and differences in transport K_m have only recently been observed with the discovery of the multiple glucose transporter isoforms with their own respective K_m values, the problem of altered or regulated intrinsic activities appears to reside solely in the turnover number. Initially, the problem arose because glucose transport activity in the intact adipose cell could be stimulated by 30–40-fold while the number of glucose transporters in the plasma membrane fraction was increased at most by up to ten fold.

Two studies of altered metabolic states by Kahn and colleagues in our laboratory at the NIH also suggested regulation of glucose transporter intrinsic activity: insulin treatment of streptozotocin-diabetic rats and refeeding of fasted rats [26,32]. Both of these studies observed transient hyperelevations in insulin-stimulated glucose transport activity during the first few days of treatment which were not accompanied by anything more than restoration of glucose transporter number back to the normal levels, either in the intracellular pool in the basal state or in the plasma membranes in the insulin-stimulated state. In an apparent extension of the concept of regulated intrinsic activity, Smith et al. [33], Kuroda et al. [34], and Joost et al. [35] clearly demonstrated that hormones and agents which activate Gs such as isoproterenol, ACTH, and glucagon inhibited insulin-stimulated glucose transport activity in the isolated rat adipose cell without a corresponding reduction in the number of glucose transporters in the plasma membranes. Early studies of the regulation of glucose transport by insulin in skeletal muscle also contributed heavily to the concept of regulated glucose transporter intrinsic activity because of even larger disparities between glucose transport activity in intact muscles and glucose transporter number in the plasma membrane fraction. Indeed, in studies in which glucose transport activity and glucose transporter number were both measured in the same isolated plasma membrane preparations of basal and insulin-stimulated skeletal muscle, the ratio of activity to number was reported to increase with insulin treatment [36]. As will be described shortly, however, this concept is no longer tenable.

Evidence for Multiple Glucose Transporters

Also during the early period of the development of the translocation hypothesis it was assumed that only one glucose transporter species existed: that readily found in the human erythrocyte plasma membrane. All of the initial glucose transport characterization studies took advantage of the high concentration of glucose transport protein in this cell type and membrane preparation, and red cell ghosts served as the source for purification of this glucose transporter. Antisera were also prepared against this protein, and the combination of partial sequences and antisera provided the basis for the cloning of the human erythrocyte glucose transporter by Mueckler et al. [37] from HepG2 cells. Because the cloning was achieved using this hepatoma cell line which expressed high levels of the same transporter found in the human erythrocyte, the concept of one glucose transporter species continued unabated although early indications of the family of glucose transporter isoforms were already present. In hindsight, it is now clear that reports from our laboratory [38] and Lienhard et al. [39] missed very significant signs of the presence of more than one glucose transporter protein in the rat adipose cell. In these studies, the translocation of glucose transporters was demonstrated by Western blotting for the first time using antisera prepared against the purified RBC glucose transporter. In the same plasma membrane preparations showing a five fold increase in glucose transporter number as assessed by cytochalasin B in response to insulin, the signals by Western blotting increased only two fold. In the same intracellular membranes showing a 50% decrease in cytochalasin B binding sites, Western blotting signals almost disappeared.

In an attempt to carry out immunocytochemical studies at the electron microscope level of the translocation of glucose transporters in collaboration with Geuze and colleagues, our laboratory provided fixed rat adipose cells and antiserum against the human erythrocyte glucose transporter, with little success. Lienhard and colleagues in collaboration with the same group used 3T3-L1 adipocytes and an affinity-purified preparation of a similar antiserum; they succeded with excellent electron micrographs clearly demonstrating what we now know as the translocation of GLUT1 [40]. While the obvious retrospective conclusion of these disparities was totally missed, a study by Oka et al. [41] compared the efficiency of immunoprecipitation of [³H]cytochalasin B-labelled glucose transporters from human erythrocytes and rat adipose cells with antiserum against the purified human RBC glucose transporter; the low efficiency of immunoprecipitation from the latter cells compared to the former was indicative of the presence a second labelled protein in the adipose cell. Clear recognition of the presence of two glucose transporter isoforms in the rat adipose cell came with the preparation of the 1F8 monoclonal antibody to a protein in a preparation of intracellular membranes by James et al. [42]. Shortly thereafter, five independent laboratories cloned what was known then as the "insulin responsive glucose transporter" to distinguish it from the "HepG2/human erythrocyte glucose transporter". Subsequent cloning studies ultimately led to identification of the remainder of the currently recognized members of the facilitated diffusion glucose transporter family [3].

Regulation of Glucose Transporter Expression

Subsequent to the cloning of GLUT1 and especially GLUT4, an extensive literature began to appear in which the altered metabolic states previously studied using the cytochalasin B binding technique to quantitate glucose transporter number were revisited using the new anti-GLUT1 and anti-GLUT4 antisera [43]. The changes in the levels of GLUT4 were by and large shown to parallel directly the previously reported changes in glucose transporter level assessed by cytochalasin B binding with both insulin resistance and insulin hyper-responsiveness. Surprisingly, GLUT1 levels changed very little. However, given the relatively low abundance of GLUT1 compared to GLUT4, the lack of change in GLUT1 would appear to have little physiological significance relative to the changes in GLUT4. As will be discussed below, detailed quantitative studies of cell surface GLUT1 and GLUT4 in intact adipose cells clearly demonstrate that GLUT4 accounts for most of the glucose uptake in highly insulin responsive cells even in the basal state. Nevertheless, if GLUT4 levels were sufficiently low, as for example in the recently reported GLUT4 knockout mouse or in markedly insulin-resistant human adipose cells in patients with NIDDM, then the unchanged GLUT1 levels might assume a much more significant role.

The area in which the cloning and identification of the individual GLUT1 and GLUT4 glucose transporter isoforms makes the major contribution is understanding the mechanisms through which the levels of these proteins are regulated [43]. Because the half-lives of these proteins appear to be quite long (perhaps 24 h or more), assessment of the mRNA levels without simultaneous measurement of the protein levels is of little value; direct correspondence between the two is not assured. Furthermore, evidence from studies of cultured cells clearly indicates that long-term regulation of glucose transporter protein occurs at both the gene expression and protein turnover levels. However, important studies of the regulation of glucose transporter gene expression are in progress with respect to identification of both the factors which influence gene expression and the promoter elements in the genes which control gene transcription. Before a complete understanding of GLUT4 translocation is in hand, we will also have to identify the cellular machinery which carries out the subcellular recycling of glucose transporters and how the expression of its components is regulated.

Virtually from the time of the discovery of more than one glucose transporter isoform, GLUT1 has been described as the constitutive isoform because of its wide spread expression in all tissues and cells examined. GLUT1 is particularly highly expressed in cells in tissue culture presumably because these cells derive most of their metabolic energy from glycolysis and thus require a continuous, large flux of glucose. Even in isolated rat adipose cells maintained in primary culture as studied in our laboratory at the NIH, the GLUT4 mRNA levels rapidly decline (within 6 h) and the GLUT4 protein levels gradually disappear (over 3–5 d). While GLUT1 protein levels increase slowly over the same period, a dramatic shift in their subcellular distribution occurs from a ratio of ~20% plasma membranes/~80% intracellular membranes, to one in which virtually all are found in the plasma membranes. Interestingly, the basal GLUT4 subcellular distribution does exactly the same thing; basal glucose transport activity gradually increases and the insulin response decreases, the latter even more rapidly than the rate of decrease in total GLUT4 as GLUT4 spontaneously redistributes to the plasma membranes. Somehow, however, the concept of GLUT1 as the constitutive glucose transporter has been translated into the concept that GLUT1 is the basal glucose transporter in insulin responsive cells. It is now clear that whether GLUT1 can or cannot be translocated depends directly on the presence or absence of the appropriate subcellular machinery for sequestering GLUT1 in a mobilizable intracellular pool. In some cells, an intracellular pool of GLUT1 appears to exist in order that those cells can respond to stimuli other than insulin, for example, in polymorphonuclear leukocytes responding to select chemotactic stimuli [44].

Photoaffinity Labelling of Glucose Transporters

Early Development

Definitive quantitative data regarding the relationship between glucose transport activity and the number and isoforms of glucose transporters only became available with the reporting by Holman and colleagues of the synthesis of an impermeant bismannose derivative capable of specifically labelling glucose transporters on their exofacial domains [45]. Holman et al. [46] had previously described a series of sugar derivatives which could be used to photolabel the human erythrocyte glucose transporter long before it was identified as GLUT1, but the affinities of these compounds were too low to be useful in labelling the relatively small numbers of glucose transporters in other cell types. The bis-mannose compound proved to have a sufficiently high affinity which together with its impermeance, provided a means for quantitating specifically those glucose transporters on the extracellular surface of intact cells. Holman et al. ultimately replaced the original azido photoreactive group with a carbene precursor diazerine group so as to improve the reactivity and specificity of the photoreaction with the glucose transporter [47]. Nevertheless, a protein in the rat adipose cell plasma membrane with an M_r of $\sim 70\,kDa$ but otherwise unidentified, is strongly labelled by the bis-mannose photolabel, although this labelling is not inhibited by competitive inhibitors of glucose transporter photolabelling such as glucose and cytochalasin B. Quantitation of glucose transporter number must, therefore, be carried out by a combination of photolabelling and immunoprecipitation.

Intrinsic Activity Revisited

ATB-BMPA labelling of several glucose transporter isoforms has been well characterized including GLUT1, GLUT3, and GLUT4 [6]. GLUT5, the "fructose transporter" is not labelled, an observation which is consistent with the non-glucose substrate profile of this glucose transporter isoform. Of considerable interest are the relatively similar affinities of GLUTs 1, 3, and 4 for the bis-mannose despite a much broader range of K_ms. This suggests that substrate binding from the exofacial direction is quite similar among the three glucose transporter isoforms and that the transport K_ms are not necessarily only indicative of substrate binding affinity.

Indeed, Holman and colleagues recently described an elegant technique for determining the relative contributions of more than one glucose transporter isoform to glucose transport in a particular cell type by combining photoaffinity labelling with [^{3}H]ATB-BMPA, competitive inhibition of photolabelling with unlabelled ATB-BMPA and one or more transported substrates with known K_ms, and immunoprecipitation individually of the various glucose transporter isoforms [48]. Competitive inhibition of photolabelling with unlabelled bis-mannose permits determination of the relative (actually absolute) number of each glucose transporter isoform while competitive inhibition of photolabelling with transported substrates of known K_ms permits assessment of the relative turnover numbers. It is exactly this approach which has clearly demonstrated that GLUT4 exhibits approximately three times the intrinsic activity (turnover number/K_m) that GLUT1 exhibits; given the roughly three fold lower K_m for GLUT4 than GLUT1, the turnover numbers for the two glucose transporter isoforms must be roughly equivalent. Thus, even the low levels of expression of GLUT4 on the surface of the highly insulin responsive rat adipose cell in the basal state are sufficient to account for most of the basal glucose transport activity.

Cell Surface Glucose Transporters

Relationship to Glucose Transport Activity

The development of the bis-mannose photolabelling technique led to two fundamental breakthroughs in further elucidating the glucose transporter translocation process. First, the widespread suggestion that regulation of glucose transport activity occurs at the level of glucose transporter intrinsic activity as well as subcellular distribution could be directly addressed by comparing glucose transport measurements with quantitative determinations of cell surface glucose transporter number [47,48]. In virtually every experimental preparation tested, the two parameters matched as exactly as experimental variation would allow. Initially, this direct relationship was clearly demonstrated in the isolated rat adipose cell comparing the rates of 3-O-methylglucose transport with the numbers of GLUT1 and GLUT4 on the cell surface in the basal and insulin-stimulated states. Despite the presence of similar amounts of GLUT1 and GLUT4 on the cell surface in the basal state, glucose transport activity and the cell surface number of GLUT4 both increased by ~20-fold in response to insulin while the cell surface number of GLUT1 increased by only ~four fold.

This same report also explained the apparent discrepancy between a small increase in glucose transport activity and a much larger increase in plasma membrane glucose transporter number as assessed by subcellular fractionation and cytochalasin B binding in response to activators of protein kinase C. The latter stimulate the translocations of GLUT1 and GLUT4 both roughly three fold while insulin stimulates GLUT4 translocation ~20-fold. Both glucose transporter isoforms bind cytochalasin B in a similar fashion, but GLUT1 has the three-fold lower intrinsic activity. Thus, insulin's stimulatory action on glucose transport in the adipose cell can be almost fully explained by the appearance of functional GLUT4 on the cell surface. Because no one has ever assessed the activity state of glucose transporters in the intracellular membranes without some intervening procedure such as reconstitution, a possible effect of insulin on the intrinsic activity of GLUT4 during its translocation to the cell surface remains to be clarified; nevertheless, reconstituted intracellular GLUT4 has the same transport characteristics as GLUT4 reconstituted from the plasma membranes.

Counterregulatory Hormones

Early experiments from our laboratory clearly demonstrated that hormones and agents which activate the heterotrimeric GTP binding proteins G_s and G_i such as isoproterenol and adenosine, respectively, decrease and increase insulin-stimulated glucose transport activity through what appeared to be modulations of glucose transporter intrinsic activity [33–35]. This conclusion was again based on comparisons of glucose transport activity in intact isolated rat adipose cells with glucose transporter number in the plasma membranes measured by cytochalasin B binding. More recent measurements by Vannucci et al. [49] have taken advantage of the bis-mannose photolabel to assess cell surface GLUT4 number. The results suggest that these agents actually shift GLUT4 between an occluded state in the plasma membrane, perhaps vesicles associated with the plasma membrane during either the exocytosis leg of the translocation process or the endocytosis leg, and the fully functional state where glucose transporters are properly inserted into the plasma membrane. Thus, photolabelled cell surface GLUT4 decreases or increases in parallel with the corresponding changes in glucose transport activity. At the same time, the concentrations of GLUT4 in the plasma membranes assessed by Western blotting remain essentially unchanged. These observations are currently under further investigation in our laboratory by attempting to measure cell surface GLUT4 by galactose oxidase oxidation and $NaB[^3H]_4$ reduction, a method which is independent of substrate/GLUT4 interaction. We have hypothesized that insulin regulates the subcellular distribution of GLUT4 while G_s and G_i participate in regulating membrane fusion during exocytosis or membrane budding during endocytosis. Detailed isoproterenol and adenosine dose response curves reported by Kuroda et al. [34] further suggest that these effects are mediated directly by the G proteins themselves rather than cAMP.

Skeletal Muscle

The availability of the bis-mannose cell surface photolabel for glucose transporters has most recently been applied in examining the regulation of glucose transport in isolated rat skeletal muscle [50,51]. The relationship between glucose transport activity in intact skeletal muscles and the concentration of plasma membrane GLUT4 has been even more difficult to quantitate than that in rat adipose cells because of the considerably more difficult problems in preparing subcellular membrane fractions from muscle than adipose cells. The consistent discrepancies between the effects of insulin and contraction, or hypoxia as a model for contraction, on glucose transport activity and the plasma membrane concentration of GLUT4 have virtually all been clarified by showing that the changes in the former are paralleled by changes in cell surface GLUT4 regardless of the previously reported concentrations of GLUT4 in plasma membrane fractions [52]. This relationship alone strongly supports the concept that insulin and contraction stimulate glucose transport in skeletal muscle through a translocation of GLUT4 similar to that observed in adipose cells even though the only data available to date on the intracellular disposition of GLUT4 still depend on measurements based on subcellular fractionation. Only kinetic studies using photolabelled cell surface glucose transporters and immunocytochemical studies at the electron microscopy level as have been carried out with adipose cells will ultimately confirm glucose transporter translocation in skeletal muscle.

Subcellular Trafficking

Recycling Kinetics

The second area of investigation which has now become approachable for the first time through use of the bis-mannose photolabel is the subcellular trafficking itself of the glucose transporters [53,54]. By labelling the cell surface glucose transporters without simultaneously labelling the intracellular glucose transporters, kinetic studies can be carried out in which trafficking between the labelled and unlabelled compartments is followed. This technique works especially well with rat adipose cells. When these cells are brought to the insulin-stimulated state, approximately 45% of the cell's total GLUT4 are on the cell surface and the difference between the basal and insulin-stimulated states is large. The movement of photolabelled GLUT4 can be readily followed either in the continuous presence of insulin or following the removal of insulin. In the rat adipose cell, so little GLUT1 is present and the basal level of cell surface GLUT4 is so low that studies of the kinetics of GLUT1 trafficking in either the insulin-stimulated or basal state, and of GLUT4 trafficking in the basal state, are not feasible. Such studies have, however, been carried out in 3T3-L1 adipocytes.

The following observations have been reported: (1) Glucose transporters move physically between compartments, as previously deduced from the reciprocal changes in GLUT4 number in subcellular fractions, but not actually shown. (2) All of the cell's glucose transporters participate in the insulin response, as evidenced by the dilution of labelled cell surface glucose transporters when cells labelled in the insulin-stimulated state are returned to the basal state and then restimulated with insulin. (3) Glucose transporters are always recycling between the intracellular compartment and the cell surface, slowly in the absence of insulin and much more rapidly in the presence of insulin. GLUT1 recycles more rapidly in the basal state than GLUT4 does, while both recycle at the same rate in the presence of insulin. (4) The difference between GLUT1 and GLUT4 in the basal state lies primarily in the more rapid rate of exocytosis of the former, thus favouring a somewhat higher steady state distribution ratio of GLUT1 to the plasma membrane. (5) Insulin appears to exert its effect primarily by stimulating exocytosis, both of GLUT4 by ~ten fold and GLUT1 by ~three fold. (6) Because the rate at which insulin stimulates GLUT4 translocation from the basal state is considerably faster than the rate of GLUT4 recycling at steady state in the presence of insulin, a simple two-compartment model of GLUT4 translo-

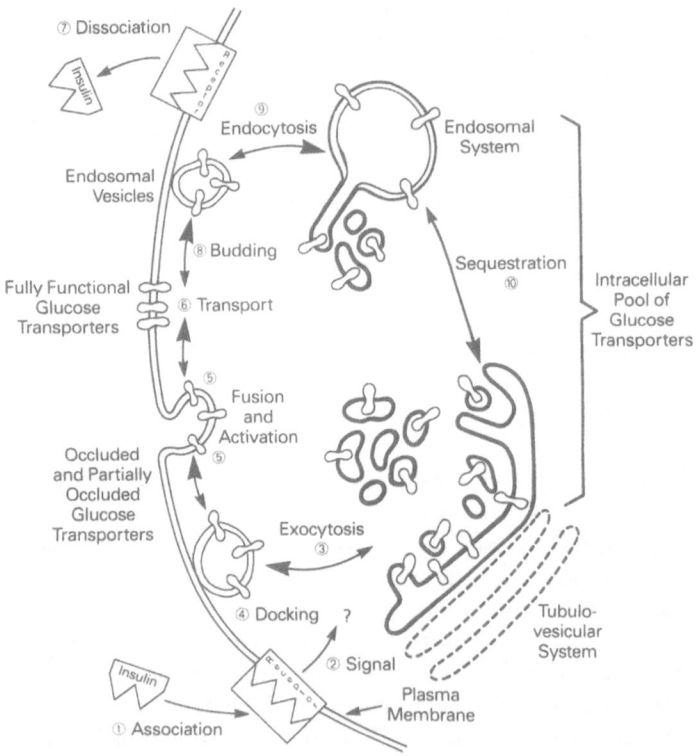

Fig. 2. Hypothetical model of insulin's action on glucose transport published in 1993. (From [55])

cation, plasma membrane and intracellular pool, is not sufficient to explain the trafficking kinetics. A minimal model which is consistent with these observations comprises two intracellular compartments in series. The basal compartment sequesters GLUT4 and is regulated by the rate of exocytosis; exit from this compartment to the plasma membrane is slow in the absence of insulin, but rapid when the rate of exocytosis is stimulated by insulin. During recycling, however, GLUT4 appears to enter another compartment from which exit back to the basal compartment becomes rate limiting when the rate of exocytosis is fast. An updated schematic representation of the translocation process is shown in Fig. 2 [55].

Many of these conclusions were simultaneously reached by Slot et al. [56] in an elegant immunohistochemical study at the electron microscope level of brown adipose cells in situ although only the steady state distributions of GLUT4 were examined. In this case, the "kinetics" were deduced because of the differential association of GLUT4 with membrane vesicles known to be in the endocytic pathway by virtue of their morphology, e.g. coated pits and associated markers such as internalized serum albumin.

Kinetic Problems

Two special problems have appeared in the literature in attempting to assess the kinetics of glucose transporter subcellular trafficking. First, adipose cells are known to exhibit elevated glucose transport activities in the apparent basal state because of rough handling, and the use of collagenase for cell preparation and serum albumin for isolated cell incubation which have not been properly screened in advance. Almost invariably, cells with decreased responses to insulin are actually partially stimulated; cells in this state would already have relatively elevated rates of exocytosis. Kinetic measurements of glucose transporter subcellular trafficking would thus underestimate the magnitude of insulin's effects on exocytosis under these conditions. The degree of underestimation would depend on the extent to which the basal rate of exocytosis had already been elevated. This problem is exemplified by a study by Jung and colleagues in which a relatively small glucose transport response to insulin appears to be due to an elevated basal glucose transport activity; these investigators concluded that insulin's effect on GLUT4 subcellular distribution was the consequence of effects of insulin of similar magnitude on exocytosis and endocytosis [57].

Second, a strong tendency is evident in the literature to assign a kinetic interpretation to steady state data. The rate at which a marker is internalized is a function simultaneously of the rates of endocytosis and exocytosis, and both the subcellular distribution of the marker and one of the rates are necessary to determine where changes may have taken place. This problem is exemplified in a study by Czech and colleagues in which only the subcellular distribution of glucose transporters was assessed; the changes observed readily support effects of insulin at either the exocytosis or endocytosis legs of the recycling process [58]. Holman and coworkers have recently described the elegant use of wortmannin, a potent and relatively specific inhibitor of PI3 kinase, to inhibit the rate of exocytosis of glucose transporters in both the basal and insulin-stimulated states in rat adipose cells and 3T3-L1 adipocytes, and even in cells in culture which are not targets of insulin action [59]. This technique permits an examination of the rates of endocytosis in a unidirectional manner and thus independently demonstrates the conclusions drawn from photolabelling and immunocytochemistry.

Concluding Remarks

The purpose of this review has been to provide an historical view of the now well supported hypothesis that insulin stimulates glucose transport through the translocation of glucose transporters, primarily GLUT4, from an intracellular storage site to the plasma membrane, as well as some anecdotes relative to this author's experience with the evolution of this hypothesis that would not appear in a stirctly scientific treatise on the subject. An obvious major area of consideration regarding insulin's action on glucose transport which has hardly been mentioned here is that of the signalling pathways which link the binding of insulin to its receptor to the translocation process. The absence of this subject is not a reflection of this author's view of its relative importance, but rather a statement of this author's lack of experience therewith. Only an investigator with in depth personal experience in the signalling field such as this author's experience with glucose transporter translocation could provide the equivalent historical view and corresponding anecdotes. Such a review would be a welcome addition to the armchair literature.

This review began with a few words about the earliest history of diabetes and the eventual discovery of insulin. This author's interest in glucose transport followed an early research experience with insulin resistance in obesity and the stimulus to identify sites in the adipose cell of the rat where insulin resistance occurs. When studies of the insulin receptor failed to explain adequately the reduced glucose metabolism response, an assay for glucose transport was developed using the nonmetabolizable pentose L-arabinose. It was only after demonstrating that glucose transport was a strategic site of insulin resistance that we realized how little was actually known about the underlying mechanism through which insulin produced this fundamental action. Thus a further understanding of diabetes and insulin resistance was the initial stimulus for the early work leading to the translocation hypothesis.

With ever increasing evidence that glucose transport is a primary site of insulin action in glucose homeostasis, it is reasonable to believe that understanding the defects in the translocation of glucose transporters in response to insulin will provide major insights into altered insulin action in diabetes, both type 1 where insulin resistance is closely related to the degree of normalization of blood glucose levels and type 2 where insulin resistance may be a fundamental part of the disease process itself. While insulin resistant glucose transport in the adipose cell appears to be quite adequately explained by changes in the number of GLUT4 molecules in the intracellular pool available for translocation in response to insulin, the focus must now shift to skeletal muscle; the latter represents the primary target cell through which insulin regulates the uptake of glucose from the blood. Little change in total muscle GLUT4 content has been observed in a variety of insulin resistant states and thus the translocation process itself becomes a prime suspect. Despite the long way still to go in understanding the details of the translocation mechanism, however, the framework has been laid and the appropriate studies are under way in laboratories throughout the world.

Acknowledgements. While the body of work reviewed here is obviously the result of contributions from many individuals with whom I have had the distinct pleasure, even honour, of associating over many years, acknowledgement of a few special relationships is nevertheless appropriate. Lester B. Salans was my mentor during the early work on glucose transporter translocation and asked me to move with him from Dartmouth Medical School to the NIH. Les' mentorship and the move to the NIH were critical to the evolution of this work. Lawrence J. Wardzala, a graduate student at Dartmouth Medical School, started the cytochalasin B binding studies in Hanover, brought them to the NIH, and carefully and creatively designed and carried out the crucial experiments leading to the translocation hypothesis. Mary Jane Zarnowski also moved to the NIH from Dartmouth Medical School, bringing with her a very special expertise in preparing and assaying glucose transport activity in isolated rat adipose cells. James E. Foley developed the first glucose transport assay in rat adipose cells in our Dartmouth laboratory also as a graduate student, but finished his graduate work before we left Hanover. In Bethesda, Ian A. Simpson joined our new laboratory at the NIH as a postdoctoral fellow and is now the laboratory's Associate Chief; Ian's superb laboratory expertise and imaginative experimental design have contributed heavily and continuously to the elaboration of the translocation hypothesis. Dena R. Yver and Steven R. Richards also joined the technical staff of our new NIH laboratory and continue today to contribute their expert technical assistance to the laboratory's experimental work.

Special thanks are offered to Jesse Roth who invited us as newcomers to the NIH to participate in Diabetes Branch activities. Jesse, together with many other Diabetes Branch investigators, supported our work under even the most discouraging circumstances with encouragement and imaginative ideas. Additional special thanks go to Tetsuro Kono who has been a particular friend and colleague from the very beginning of the concept of glucose transporter translocation. While discovery of the translocation process took place independently in our two laboratories, many of the early methods and concepts related to rat adipose cell function were the result of Tetsuro's pioneering work. Finally, the remarkable progress made in the past few years in understanding the quantitative relationship between glucose transport activity, and glucose transporter number and subcellular trafficking are the direct consequence of a now long-standing, productive, and creative collaboration with Geoffrey D. Holman in whose laboratory this review was written during my sabbatical. Geoff's innovative development of the bis-mannose photoaffinity label has not only set a quantitative standard to which all work in this field of study must now be compared, but also provided a unique approach to pursuing the mechanism of insulin's stimulatory action on glucose transport.

References

1. Papaspyros MS (1964) The history of diabetes mellitus, 2nd edn. Thieme Stuttgart
2. Banting FG, Best CH (1921) Pancreatic extracts. J Lab Clin Med 7: 464–472
3. Gould GW, Holman GD (1993) The glucose transporter family: structure, function and tissue-specific expression. Biochem J 295: 329–341
4. Craik JD, Stewart M, Cheeseman CI (1995) GLUT3 (brain-type) glucose transporter polypeptides in human blood platelets. Thromb Res 79: 461–469
5. Bell GI, Kayano I, Buse JB, Burant CF, Takeda J, Lin D, Fukumoto H (1990) Molecular biology of mammalian glucose transporters. Diabetes Care 13: 198–208
6. Wilson CM, Mitsumoto Y, Maher F, Klip A (1995) Regulation of cell surface GLUT1, GLUT3, and GLUT4 by insulin and IGF-I in L6 myotubes. FEBS Lett 368: 19–22
7. Levine R (1961) Concerning the mechanisms of insulin action. Diabetes 10: 421–431
8. Crofford OB, Renold AE (1965) Glucose uptake by incubated rat epididymal adipose tissue. Characteristics of the glucose transport system and action of insulin. J Biol Chem 240: 3237–3244
9. Rodbell M (1964) Metabolism of isolated fat cells. I. Effects of hormones on glucose metabolism and lipolysis. J Biol Chem 239: 375–480
10. Gliemann J, Rees WD (1983) The insulin-sensitive hexose transport system in adipocytes. Curr Top Membr Transp 18: 339–379

11. Martin DB, Carter JR (1970) Insulin-stimulated glucose uptake by subcellular particles from adipose tissue cells. Science 167: 873–874
12. Wardzala LJ, Cushman SW, Salans LB (1978) Mechanism of insulin action on glucose transport in the isolated rat adipose cell. Enhancement of the number of functional transport systems. J Biol Chem 253: 8002–8005
13. Lin S, Spudich JA (1874) Biochemical studies of the mode of action of cytochalasin B. Cytochalasin B binding to red cell membrane in relation to glucose transport. J Biol Chem 249: 5778–5783
14. Cushman SW, Wardzala LJ (1980) Potential mechanism of insulin action on glucose transport in the isolated rat adipose cell. Apparent translocation of intracellular transport systems to the plasma membrane. J Biol Chem 255: 4758–4762
15. Suzuki K, Kono T (1980) Evidence that insulin causes translocation of glucose transport activity to the plasma membrane from an intracellular storage site. Proc Natl Acad Sci USA 77: 2542–2545
16. Karnieli E, Zarnowski MJ, Hissin PJ, Simpson IA, Salans LB, Cushman SW (1981) Insulin-stimulated translocation of glucose transport systems in the isolated rat adipose cell. Time course, reversal, insulin-concentration-dependency and relationship to glucose transport activity. J Biol Chem 256: 4772–4777
17. Simpson IA, Cushman SW (1985) Hormonal regulation of mammalian glucose transport. Annu Rev Biochem 55: 1059–1089
18. Kono T, Barham FW (1971) The relationship between the insulin-binding capacity of fat cells and the cellular response to insulin. Studies with intact and trypsin-treated fat cells. J Biol Chem 246: 6210–6216
19. Wardzala LJ, Simpson IA, Rechler MM, Cushman SW (1984) Potential mechanism of the stimulatory action of insulin on insulin-like growth factor II binding to the isolated rat adipose cell. Apparent redistribution of receptors cycling between a large intracellular pool and the plasma membrane. J Biol Chem 259: 8378–8383
20. Lobel P, Dahms NM, Breitmeyer J, Chirgwin JM, Kornfield S (1987) Cloning of the bovine 215-kDa cation-independent mannose 6-phosphate receptor. Proc Natl Acad Sci USA 84: 2233–2237
21. Sonne O, Simpson IA (1984) Internalization of insulin and its receptor in the isolated rat adipose cell: time course and insulin concentration dependency. Biochim Biophys Acta 804: 404–413
22. Wolosin JM, Forte JG (1981) Changes in the membrane environment of the (K^++H^+)-ATPase following stimulation of the gastric oxyntic cell. J Biol Chem 256: 3149–3152
23. Wade JB, Stetson DL, Lewis SA (1981) ADH action: evidence for a membrane shuttle mechanism. Ann NY Acad Sci 372: 106–117
24. Karnieli E, Hissin PJ, Simpson IA, Salans LB, Cushman SW (1981) A possible mechanism of insulin resistance in the rat adipose cell in streptozotocin-induced diabetes mellitus. Depletion of intracellular glucose transport systems. J Clin Invest 68: 811–814
25. Hissin PJ, Karnieli E, Simpson IA, Salans LB, Cushman SW (1982) A possible mechanism of insulin resistance in the rat adipose cell with high fat/low carbohydrate feeding. Depletion of intracellular glucose transport systems. Diabetes 31: 589–592
26. Kahn BB, Simpson IA, Cushman SW (1988) Divergent mechanisms for the insulin resistant and hyperresponsive glucose transport in adipose cells from fasted and refed rats. Alterations in both glucose transporter number and intrinsic activity. J Clin Invest 82: 691–699
27. Karnieli E, Chernow R, Hissin PJ, Simpson IA, Foley JE (1986) Insulin stimulates glucose transport in isolated human adipose cells through the translocation of intracellular glucose transporters to the plasma membrane. Horm Metabol Res 18: 860–861
28. Horuk R, Rodbell M, Cushman SW, Wardzala LJ (1983) Proposed mechanism of insulin-resistant glucose transport in the isolated guinea pig adipocyte. Small intracellular pool of glucose transporters. J Biol Chem 258: 7425–7429
29. Kahn BB, Horton ES, Cushman SW (1987) Mechanism for enhanced glucose transport response to insulin in adipose cells from chronically hyperinsulinemic rats. Increased translocation of glucose transporters from an enlarged intracellular pool. J Clin Invest 79: 853–858
30. Hissin PJ, Foley JE, Wardzala LJ, Karnieli E, Simpson IA, Salans LB, Cushman SW (1982) Mechanism of insulin resistant glucose transport activity in the enlarged adipose cell of the aged, obese rat. Relative depletion of intracellular glucose transport systems. J Clin Invest 70: 780–790
31. Guerre-Millo M, Lavau M, Horne JS, Wardzala LJ (1985) Proposed mechanism for increased insulin-mediated glucose transport in adipose cells from young, obese Zucker rats. J Biol Chem 260: 2197–2201

32. Kahn BB, Cushman SW (1987) Mechanism for markedly hyperresponsive insulin-stimulated glucose transport activity in adipose cells from insulin-treated streptozotocin diabetic rats. Evidence for increased glucose transporter intrinsic activity. J Biol Chem 262: 5118–5124
33. Smith U, Kuroda M, Simpson IA (1984) Counterregulation of insulin-stimulated glucose transport by catecholamines in the isolated rat adipose cell. J Biol Chem 259: 8758–8763
34. Kuroda M, Honnor RC, Cushman SW, Londos C, Simpson IA (1987) Regulation of insulin-stimulated glucose transport in the isolated rat adipocyte. cAMP-independent effects of lipolytic and antilipolytic agents. J Biol Chem 262: 245–253
35. Joost HG, Weber TM, Cushman SW, Simpson IA (1986) Insulin-stimulated glucose transport in rat adipose cells. Modulation of transporter intrinsic activity by isoproterenol and adenosine. J Biol Chem 261: 10033–10036
36. Goodyear LJ, Hirshman MF, Smith RJ, Horton ES (1991) Glucose transporter number, activity, and isoform content in plasma membranes of red and white skeletal muscle. Am J Physiol 261: E556–E561
37. Mueckler M, Caruso C, Baldwin SA, Panico M, Blench I, Morris HR, Allard WJ, Lienhard GE, Lodish HF (1985) Sequence and structure of a human glucose transporter. Science 229: 941–945
38. Wheeler TJ, Simpson IA, Sogin DC, Hinkle PC, Cushman SW (1982) Detection of the rat adipose cell glucose transporter with antibody against the human red cell glucose transporter. Biochem Biophys Res Commun 105: 89–95
39. Lienhard GE, Kin HK, Ransome KJ, Gorga JC (1982) Immunological identification of an insulin-responsive glucose transporter. Biochem Biophys Res Commun 105: 1150–1156
40. Blok J, Gibbs EM, Lienhard GE, Slot JW, Geuze HJ (1988) Insulin-induced translocation of glucose transporters from post-Golgi compartments to the plasma membrane of 3T3-L1 adipocytes. J Cell Biol 106: 69–76
41. Oka Y, Asano T, Shibashi Y, Kasuga M, Kanasawa Y, Takaku F (1988) Studies with antipeptide antibody suggest the presence of at least two types of glucose transporters in rat brain and adipocyte. J Biol Chem 263: 13432–13439
42. James DE, Brown R, Navarro J, Pilch PF (1988) Insulin-regulatable tissues express a unique insulin-sensitive glucose transport protein. Nature 333: 183–185
43. Abel ED, Sheperd P, Kahn BB (1996) Glucose transporters and pathophysiologic states. In: LeRoith D, Taylor SI, Olefsky JM (eds) Diabetes mellitus. Lippincott-Raven, Philadelphia, pp 530–543
44. Okuno Y, Gliemann J (1988) Effect of chemotactic factors on hexose transport in polymorphonuclear leucocytes. Biochim Biophys Acta 941: 157–164
45. Clark AE, Holman GD (1990) Exofacial photolabelling of the human erythroctye glucose transporter with an azitrifluoroethylbenzoyl-substituted bismannose. Biochem J 269: 615–622
46. Holman GD, Karim AR, Karim B (1988) Photolabelling of erythrocyte and adipocyte hexose transporters using a benzophenone derivative of bis (D-mannose). Biochim Biophys Acta 946: 75–84
47. Holman GD, Kozka IJ, Clark AE, Flower CJ, Saltis J, Habberfield AD, Simpson IA, Cushman SW (1990) Cell surface labelling of glucose transporter isoform GLUT4 by bis-mannose photolabel in rat adipose cells. Correlation with stimulation of glucose transport by insulin and phorbol ester. J Biol Chem 265: 18172–18179
48. Palfreyman RW, Clark AE, Denton RM, Holman GD, Kozka IJ (1992) Kinetic resolution of the separate GLUT1 and GLUT4 glucose transport activities in 3T3-L1 cells. Biochem J 284: 275–281
49. Vannucci SJ, Nishimura H, Satoh S, Cushman SW, Holman GD, Simpson IA (1992) Cell surface accessibility of GLUT4 glucose transporters in insulin-stimulated rat adipose cells. Modulation by isoprenaline and adenosine. Biochem J 288: 325–330
50. Wilson CM, Cushman SW (1994) Insulin stimulation of glucose transport activity in rat skeletal muscle. Increase in cell surface GLUT4 as assessed by photolabelling. Biochem J 299: 755–759
51. Lund S, Holman GD, Schmitz O, Pedersen O (1993) GLUT4 content in the plasma membrane of rat skeletal muscle: comparative studies of the subcellular fractionation method and the exofacial photolabelling technique using ATB-BMPA. FEBS Lett 330: 312–318
52. Lund S, Holman GD, Schmitz O, Pedersen O (1995) Contraction stimulates translocation of glucose transporter GLUT4 in skeletal muscle through a mechanism distinct from that of insulin. Proc Natl Acad Sci USA 92: 5817–5821

53. Holman GD, Leggio LL, Cushman SW (1994) Insulin-stimulated GLUT4 glucose transporter recycling. A problem in membrane protein subcellular trafficking through multiple pools. J Biol Chem 269: 17516–17524

54. Holman GD, Cushman SW (1994) Subcellular localization and trafficking of the GLUT4 glucose tranporter isoform in insulin-reponsive cells. Bioessays 16: 753–759

55. Satoh S, Nishimura H, Clark AE, Kozka IJ, Vannucci SJ, Simpson IA, Quon MJ, Cushman SW, Holman GD (1993) Use of bis-mannose photolabel to elucidate insulin-regulated GLUT4 subcellular trafficking kinetics in rat adipose cells. Evidence that exocytosis is a critical site of hormone action. J Biol Chem 268: 17820–17829

56. Slot JW, Geuze HJ, Gigengack S, Lienhard GE, James DE (1991) Immunolocalization of the insulin regulatable glucose transporter in brown adipose tissue of the rat. J Cell Biol 113: 123–135

57. Jhun BH, Rampai AL, Liu M, Lachaal M, Jung CY (1992) Effects of insulin on steady state kinetics of GLUT4 subcellular distribution in rat adipocytes. Evidence of constitutive GLUT4 recycling. J Biol Chem 267: 17710–17715

58. Czech MP, Buxton JM (1993) Insulin action on the internalization of the GLUT4 glucose transporter in isolated rat adipocytes. J Biol Chem 268: 9187–9190

59. Yang J, Clarke JF, Ester CJ, Young PW, Kasuga M, Holman GD (1996) Phosphatidylinositol 3-kinase acts at an intracellular membrane site to enhance GLUT4 exocytosis in 3T3-L1 cells. Biochem J 313: 125–131

Insulin-like Growth Factor: Endocrine and Autocrine/Paracrine Implications and Relations to Diabetes Mellitus

E.R. FROESCH

Homologies Between and Functional Relevance of IGFs and Insulin and Their Respective Receptors

Insulin-like growth factors I and II (IGF I and IGF II) were characterized as peptides of the insulin family by Rinderknecht [1] and Humbel [2] (Fig. 1). When this was achieved, the term insulin-like became acceptable since the homology between IGF I and IGF II, on the one hand, and insulin, on the other, was found to be on the order of 45%. The IGFs consist of an A- and a B-chain, a conserved connecting peptide which is shorter than the C-peptide of insulin and a D-domain. Many years before the chemical characterization of IGFs, biological features of impure preparations of IGFs were clearly shown to have insulin-like properties. IGFs, then still termed nonsuppressible insulin-like activity, stimulated glucose uptake and lipogenesis of adipose tissue, inhibited lipolysis of adipose tissue, stimulated glucose and amino acid transport into the diaphragm, skeletal and heart muscle and caused hypoglycemia when injected intravenously. Binding experiments with adipose tissue, fat cells, heart and other cells clearly showed that the IGF preparations had an affinity for the insulin receptor and that they had clear insulin-like and mitogenic effects on some tissues at concentrations below those necessary for the displacement of insulin from the insulin receptor [3]. Later, a type 1 IGF receptor was detected which also exhibits a considerable degree of homology with the insulin receptor and also consists of two α- and two β- subunits with a tyrosine kinase as a phosphorylation-activating mechanism [4]. Type 1 IGF receptors are present on almost all undifferentiated and differentiated cells. Since insulin at high concentrations cross-reacts with type 1 receptors, it is difficult to distinguish the components activated in the insulin signal transduction pathway through the insulin receptor from those activated via the type 1 receptor (Fig. 2). It is likely that some mechanisms are similar and others different. The major difference in the way these two peptides are signalling may actually be due to the different dissociation rates of the signalling peptide from the respective receptor. Within the past few years, insulin analogues which are more hydrophilic than insulin itself have been synthesized in order to obtain insulin preparations which are resorbed faster from the subcutaneous tissue than insulin and, therefore, could compensate faster and better for hyperglycaemia resulting from carbohydrate-rich meals. These hydrophilic insulin analogues have a somewhat higher affinity for the type 1 IGF receptor than insulin itself and, in addition, appear to have a slower dissociation rate from the insulin receptor. In terms of differences between signal transduction through the type 1 and the insulin receptor, these hydrophilic insulins were even more mitogenic than could be accounted for by cross-reaction with the type

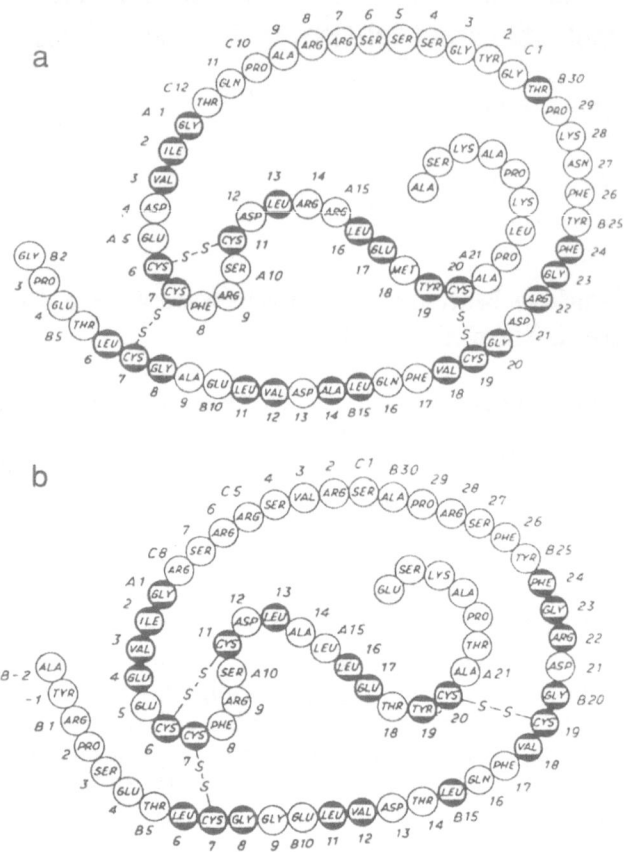

Fig. 1a,b. Primary structure of insulin-like growth factor I (**a**) and II (**b**) (from Rinderknecht and Humbel [1] and from Humbel [2])

1 receptor [5]. Therefore, even if some of the intracellular signals emitted from the insulin receptor and from the type 1 IGF receptor may be similar, major differences could be due to different dissociation rates of the peptides from their respective receptors (Table 1).

Insulin has always been considered to be a mitogen when present in high concentrations and has been used to stimulate replication in cell culture systems. It is probable that these insulin effects came about via signalling through the type 1 and not through the insulin receptor because the same mitogenic effects are achieved with small physiologic concentrations of IGF I. Therefore, one can conclude that insulin is not a major hormone with mitogenic and differentiation-enhancing activities in the concentrations in which it is present in serum. It is the major blood glucose-lowering and glucose homeostasis-maintaining hormone but not a major mitogen. On the other hand, IGF I, be it as an endocrine hormone or as an autocrine/paracrine cytokine, is a very important stimulator of cell and tissue differentiation and of

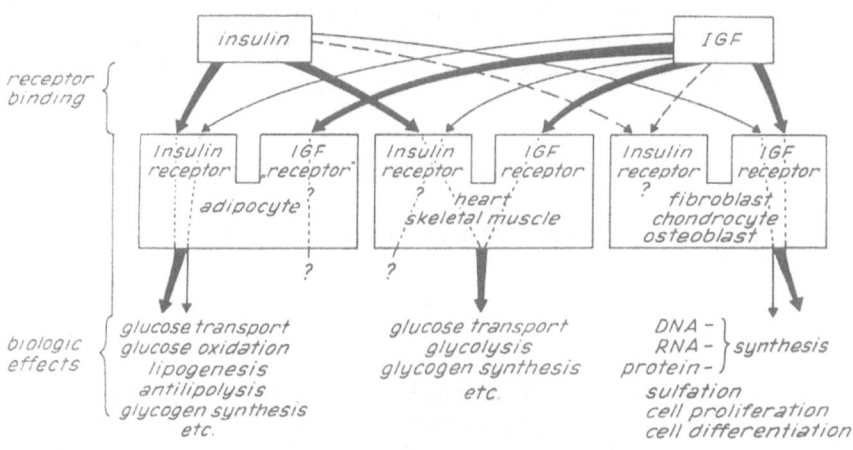

Fig. 2. Complementary anabolic action of insulin and insulin-like growth factor (*IGF*)

Table 1. Major biological differences between insulin and IGF I

	Insulin	IGF I
Mode of secretion	Emeiocytosis, regulated bursts, only β-cells	Constitutive, slow, hepatocytes (endocrine), other cells (autocrine/paracrine)
Regulation of secretion	Mostly plasma glucose level	Nutrition, growth hormone, insulin
Plasma level	0.5 µg/ml	100–250 ng/ml
$\frac{T}{2}$	5–10 minutes	12–16 h
Binding proteins	None	Six, of which three major ones in serum
Receptors	Insulin receptor	Type 1 IGF receptor
Receptor localization	All tissues	Most tissues, except liver and adipose tissue
Mode of action	Substrate homeostasis, fast	Differentiation and growth processes, slow
Hormone deficiency	Diabetes, ketoacidosis	Dwarfism (GH-insensitive)
Hormone excess	Hyperinsulinism, hypoglycaemia	Extrapancreatic tumour hypoglycaemia ("big" IGF II)
Receptor deficiency	Insulin-resistant (type A) diabetes	Unknown, intrauterine death?

IGF, insulin-like growth factor; GH, growth hormone.

growth [6–8]. Early in evolution, a precursor peptide of insulin and of the IGFs was probably responsible for all these processes, namely glucose homeostasis, anabolism, differentiation and growth. After gene duplication occurred, insulin took on the task of blood glucose regulation by second-to-second secretory pulses from the β-cells of the pancreas, whereas IGF I took on a completely different task, namely that of a

hormone and cytokine which is present in constant amounts in serum and cell surroundings, leading to proliferation and differentiation without a major role in the
acute regulation of glucose homeostasis.

Classical and Not So Classical Aspects of the Endocrine Nature of IGF I

Serum levels of IGF I are regulated mainly by three factors:

1. Growth hormone (GH)
2. Insulin
3. Nutrition

These three regulatory factors are highly interrelated (Fig. 3). Thus, food intake
leading to an increase of the glucose and amino acid concentration in serum elicits
an acute and very rapid secretion of insulin which stops glucose production by the
liver and induces uptake and storage of glucose in muscle and liver. At the same
time, insulin stops lipolysis so that the level of free fatty acids falls and glucose
temporarily becomes the main source of energy expenditure during the postprandial
period. GH is secreted in peaks during the night and in smaller peaks after protein-

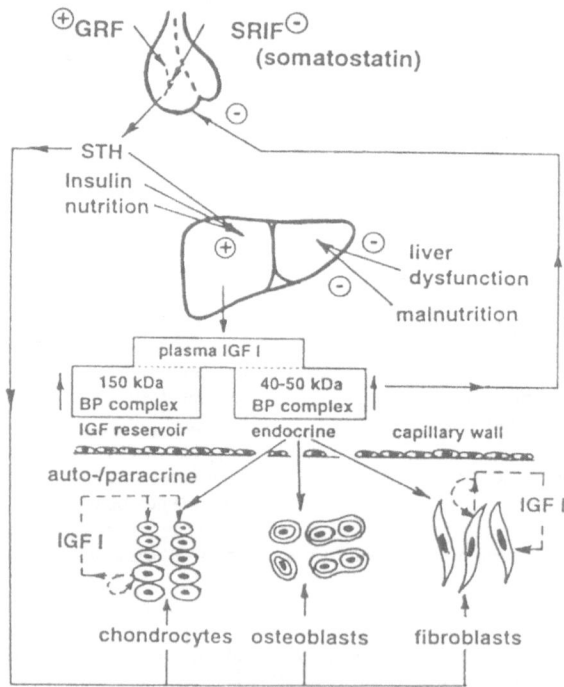

Fig. 3. Somatomedin concept (1994), *GRF*, growth hormone releasing factor; *SRIF*, somato-tropin-
releasing inhibitory factor (somatostatin); *STH*, somatropic hormone; *IGF-I*, insulin-like growth factor I; *BP*, binding protein

rich meals when amino acid levels increase. On the other hand, a rise of glycaemia suppresses GH secretion. When nutrition is inadequate, insulin secretion is decreased and GH secretion may actually increase and render the organism insulin-insensitive or insulin-resistant by increasing lipolysis and ketogenesis and turning the organism into a lipid-oxidizing machine. During prolonged fasting or in the insulin-deficient diabetic organism, IGF secretion from the liver falls and IGF I serum levels are decreased. One can conclude that GH can act on IGF synthesis in the liver only when nutrition is adequate and insulin is present. In all other situations, GH does not lead to an increase of IGF I levels. In many pathophysiological conditions in which either nutrition is inadequate or insulin lacking, GH secretion is increased, but due to the lack of insulin and of adequate nutrition GH is unable to stimulate IGF I synthesis. Therefore, GH actions are mediated by IGF I only in so far as they are anabolic, assisted by insulin and adequate food intake. In all other situations, i.e. in the presence of low IGF I concentrations, insulin deficiency and inadequate food supply or nutrient uptake by the tissues, GH stimulates lipolysis, ketogenesis and gluconeogenesis.

Metabolic Consequences of Insulin Deficiency

Insulin deficiency leads to hyperglycaemia, increased levels of free fatty acids and ketonaemia, glucosuria and finally to dehydration and death. During the development of hyperglycaemia and ketosis due to insulin deficiency, IGF I levels fall. In the case of insulin, there is no redundant hormone system which can compensate for insulin deficiency. Although IGF I mimics many of the effects of insulin in vitro and can mimic insulin effects when administered in large amounts in vivo, the endogenous IGF I levels fall in the presence of insulin deficiency and cannot compensate in any way for insulin deficiency.

IGF I Versus Insulin Deficiency

In contrast, IGF I deficiency is compatible with life and causes only minor disturbances of glucose homeostasis, but it leads to dwarfism. IGF I is the major growth-promoting hormone. One of the rare syndromes of dwarfism, namely GH-insensitive Laron dwarfism, is characterized by extremely low levels of IGF I. A dysfunction of any step between the interaction of GH with the GH receptor or any postreceptor defect, including a defect in IGF I synthesis itself, leads to so-called GH-insensitive dwarfism [9]. As neonates, these dwarfs have a tendency towards hypoglycaemia, probably because the defective GH receptors on hepatocytes and adipocytes cannot respond to GH, so that gluconeogenesis and lipolysis are not stimulated by GH and the availability of glucose and free fatty acids as substrates is reduced. During fasting the organism of the dwarf continues to depend more on glucose oxidation than a normally functioning organism. Replacement therapy with recombinant human IGF I (rhIGF I) restores growth in these dwarfs [10].

In this context, it is of great interest that replacement therapy with IGF I in insulin-deficient streptozotocin-diabetic rats does not restore glucose homeostasis but retains clear-cut anabolic properties restoring growth to some extent despite massive glucosuria and ketonuria [11]. These results again show that IGF I is particularly important for those cells that multiply and produce matrix proteins necessary for growth processes. Insulin treatment in these same diabetic rats does, of course, also normalize growth processes, mainly via normalization of IGF I synthesis in the liver and of IGF I serum levels. These interrelations between insulin and IGF I are, therefore, very important for the understanding of overall metabolism under physiologic and pathophysiologic conditions. Insulin deficiency is fatal; IGF I deficiency causes dwarfism but is compatible with life and does not cause any major malformations.

Insulin Versus IGF I Receptor Defects

The reverse is true for respective receptor deficiencies. Type A diabetic patients lacking functional insulin receptors are hyperglycaemic but in most other respects normal. In particular, they have no major organic or mental defects and are not dwarfed. In contrast, no patient with a nonfunctional type 1 IGF receptor has as yet been described. In all likelihood, such a defect is lethal. In fact, it has been shown that type 1 IGF receptor knockout mice are not viable. Whereas insulin may exert IGF I-like effects through the type 1 IGF receptor permitting normal differentiation of tissues and development in insulin receptor-deficient patients, an IGF I receptor defect does not permit normal differentiation of cells and tissues and, therefore, is not compatible with life [12].

The Food and Famine Theory: Role of IGF I

In 1963, Zierler and Rabinowitz [13] formulated the so-called "food and famine theory" on the respective roles of insulin and GH in the diurnal metabolic rhythms in humans as follows:

"We propose that, during a day, metabolism is dominated alternately by the action of insulin, or of human GH, or of the combined effects of the two, in a three-cycle phase determined by the intake of food. Exposure to insulin in the immediate postprandial period encourages storage of carbohydrate and fat, exposure to human GH plus insulin in the delayed postprandial period encourages protein synthesis and exposure to human GH in the remote postprandial phase encourages mobilization and peripheral oxidation of fat and the translocation of glucose into muscle and adipose tissue."

In the same year of the publication of this hypothesis, nonsuppressible insulin-like activity (NSILA) was described [14]. NSILA was later purified from plasma and characterized as insulin-like growth factor I and II. As already mentioned, serum levels of IGF I depend on nutrition, GH and insulin. Some of the anabolic effects of GH, insulin and IGF I are enhanced when these hormones act in concert. In the three-cycle phase of human metabolism determined by food intake, insulin is the major player in the immediate postprandial period. Insulin enhances glucose and amino acid uptake and storage by cells as well as lipid synthesis and storage. IGF I induces sensitivity towards the action of insulin and suppresses proteolysis. In the early postprandial period, all

Table 2. Endocrine and metabolic effects of administration of growth hormone versus insulin-like growth factor-I or both together

	Endocrine effects of GH	IGF I	
Insulin secretion	↑	↓	opposite
Insulin sensitivity	↓	↑	
	Metabolic effects		
Glycaemia	↑	↓	opposite
Glucose tolerance	↓	↑	
FFA	↑	↑	
β-OH-butyrate	↑	↑	
Lipid oxidation	↑	↑	
Energy expenditure	↑	↑	additive
Protein oxidation	↓	↓	
Proteolysis	↓	↓	
Amino acid levels	↓	↓	
Glucose utilization	↓	→	

FFA, free fatty acids; GH, growth hormone; IGF I, insulin-like growth factor-I.

three hormones promote anabolism with IGF I as the main modulator in mediating and/or supporting the effects of both insulin and GH. In the remote postprandial period, IGF I attenuates the induction of insulin resistance induced by GH and simultaneously promotes lipolysis by reducing insulin secretion, thus allowing for increased combustion of lipids. In addition, IGF I inhibits proteolysis and together with GH decreases protein oxidation (Table 2). For the discussion of the functions of IGF I in the food and famine theory, it is important to remember that two extremely insulin-sensitive tissues, namely the liver and adipose tissue, do not express type 1 IGF receptors. Since IGF I decreases insulin secretion and does not by itself act on the liver and on adipose tissue, it is this anabolic hormone which will switch the organism from the use of glucose to the use of free fatty acids because lipolysis is no longer inhibited by insulin.

NIDDM, Insulin Resistance, IGF I, and the Carnivore Connection

It has lately been hypothesized that the non-insulin-dependent diabetes mellitus (NIDDM) explosion in certain ethnic groups such as American Indians, Polynesians and other populations who have lived mostly on protein because they had not developed agriculture, became diabetic when suddenly confronted with an "omnivorous dietary regime", i.e. junk food of the Western civilization (Fig. 4). They had not enough time to adapt to the sudden surplus of carbohydrates. The theory says that during the many centuries during which most of the food consisted of protein and fat, blood glucose had to be synthesized by the liver via gluconeogenesis from amino acids; insulin requirement was low, and genes related to insulin resistance had a positive effect on survival [15]. When these human beings were all of a sudden

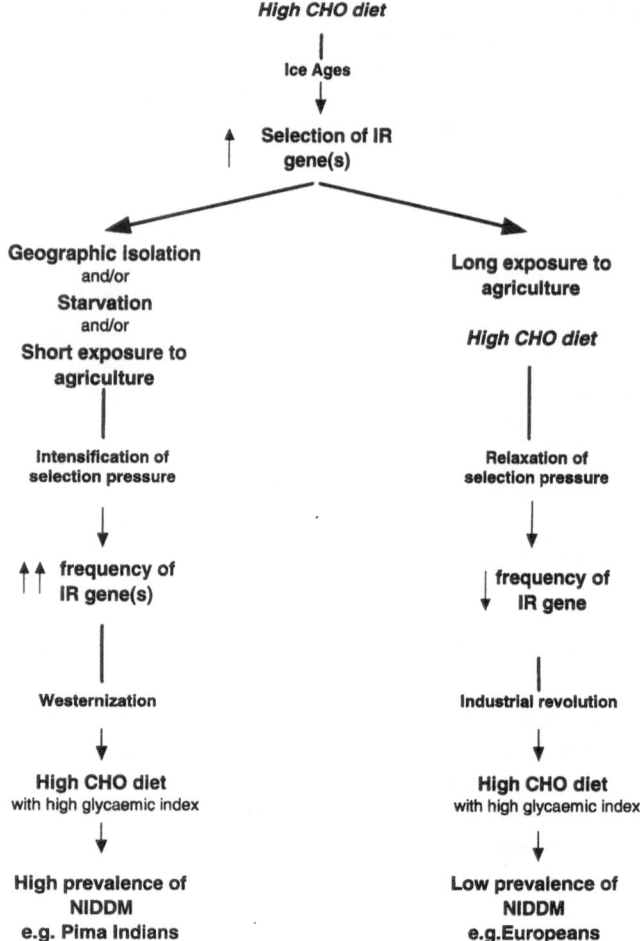

Fig. 4. The carnivore connection. *IR*, insulin resistance; *NIDDM*, non-insulin-dependent diabetes mellitus. Adapted from [15]

exposed to Western food in which protein is mixed with a lot of fat and carbohydrates, they were incapable of mastering the glucose load, became hyperglycaemic, hyper-insulinaemic, developed obesity and, later on, NIDDM. This theory is compatible with the thrifty genotype of NIDDM, postulated by Neel [16]. Many isolated ethnic populations lived on hunting and fishing and, therefore, ate mostly protein and fat. Food supply depended on the success of hunting and fishing and times of food surplus alternated with long periods of famine. During fasting, insulin levels were low and insulin resistance developed. GH in this situation was important for the mobilization of fat from adipose tissue, and IGF I helped to maintain muscle protein by inhibiting muscle proteolysis. Therefore, a certain degree of insulin resistance together with fat accumulation during times of food surplus was ideal for survival during a famine. These favourable survival factors, combined with a certain degree of insulin resistance, became hazardous in our Western society where there is a constant surplus of food which increases the probability to develop NIDDM. It is conceivable that a special setting of the GH-insulin-IGF I axis is responsible for relative insulin resistance leading to NIDDM. As we will see later, the administration of IGF I increases insulin sensitivity to a considerable extent, so that patients with NIDDM can be successfully treated with rhIGF I.

IGF Binding Proteins (IGFBPs)

The total concentration of IGF I in plasma in normal subjects is in the order of 25 nmol/l, i.e., about 100 times that of insulin. The molar concentration of IGF II is about three times that of IGF I so that the total concentration of IGF I + IGF II in plasma under normal circumstances is approximately 1 μg/ml in contrast to that of insulin which in the fasting state is around 0.5 ng/ml. If most of the IGF was freely available to cells, this huge quantity of IGF I and II would cause maximal glucose uptake of all insulin-sensitive tissues and inhibition of gluconeogenesis via the insulin receptor and, consequently, hypoglycaemia. As will be discussed later, extrapancreatic tumour hypoglycaemia is the example of hypoglycaemia induced by IGF.

Many years ago, it was found that most of the IGF I and II was present in a plasma fraction corresponding to a molecular mass of 150 000 daltons. When labelled IGF was added to serum and the serum subsequently subjected to gel filtration, most of the IGF I and II appeared in this 150 kDa fraction [17,18]. The same was found when labelled IGF was injected into rats or man [19]. These experiments demonstrated that IGF I and II were carried by binding proteins. In the meantime, six different binding proteins have been detected and also cloned from serum and from other body fluids. The major carrier protein of IGFs in serum is BP-3 which has a molecular mass of about 50 000 daltons. BP-3 binds IGF I and the BP-3 IGF complex then binds to an acid-labile subunit (ALS) which stems from the liver, is regulated by GH and keeps IGF I firmly bound in the 150 kDa complex. Other binding proteins present in plasma are IGFBP-2 and IGFBP-1 to which a small amount of IGFs is bound under normal circumstances [20].

In the large complex, IGF I and IGF II have a half-life between 12 and 16 h in adult humans. This slow turnover is explained by the fact that the 150 kDa complex does not cross the capillary barrier so that IGFs in this form cannot cross-react with their cellular receptors. Only a tiny portion of total IGF I and II is present as the free peptide, whereas approximately 10%–15% are bound to uncomplexed BP-3, BP-2 and, to a lesser extent, to BP-1. IGF I bound to these small molecular binding proteins has a half-life of approximately 20–30 min and can cross the capillary barrier and become available to cells [21].

Under most circumstances, the BP-3 levels correlate relatively well with total IGF I levels. However, this is only the case when enough ALS, also produced by hepatocytes and controlled by GH, is present in serum. The hepatic synthesis of IGF I appears to be under the same control as the hepatic synthesis of the ALS. Since IGF I bound to BP-3 forms a complex with ALS, BP-3 plasma levels must parallel those of IGF I. IGF I as such has a very short half-life of a few minutes, the BPs and the BP IGF complex without ALS one of 20–30 min, and the IGFBP-3 ALS complex a very much longer half-life between 12 and 16 h.

Since ALS and IGF I are under the control of GH, one would suspect that GH would not increase the availability of IGF I at the tissue level, even though IGF synthesis and IGF plasma levels are increased. Nevertheless, it is quite clear that increased levels of IGF I in the 150 kDa complex are accompanied by increased growth in puberty, acromegaly and gigantism as well as during GH therapy and that growing tissues have access to this increased concentration of IGF I present in the 150 kDa fraction in an as yet unexplained manner.

IGFs and Extrapancreatic Tumour Hypoglycaemia

IGFBP-1 concentrations in serum depend on insulin secretion. When insulin secretion is stimulated for instance by glucose, BP-1 levels fall acutely and it has been hypothesized that through this fall of BP-1 levels, IGF I may become more easily accessible to tissues so that IGF I may actually be a glucohomeostatic peptide. This hypothesis is still under debate. BP-1 is a very small fraction of the total pool of BPs in serum so that it is difficult to believe that such a small fraction of serum IGF I may be of importance for glucose homeostasis. Nevertheless, IGFs can become hypoglycaemic substances when they are produced in large amounts, such as is the case in extrapancreatic tumour hypoglycaemia. Zapf et al. [22] and Daughaday et al. [23] have shown that most of the large tumours causing hypoglycaemia are characterized by the following set of metabolic and endocrine abnormalities: hypoglycaemia, very low or unmeasurable insulin levels, markedly decreased or unmeasurable GH levels, inhibited lipolysis and glucose production from the liver. It was found that these tumours produce "big" IGF II in large amounts. "Big" IGF II is an incompletely processed IGF II which binds to binding proteins and to the type 1 IGF receptor as well as to the insulin receptor. When produced in large amounts, "big" IGF II inhibits insulin and GH secretion directly at the level of the β-cell and the somatotrophic cell. Due to the fact that GH secretion is inhibited, the liver ceases to synthesize the ALS so that "big" IGF II now is no longer bound in the 150 kDa complex but rather present as the free peptide or bound to the various non-GH-dependent IGF binding proteins in serum. These can cross the capillary barrier so that big IGF II may react with the type 1 IGF and the insulin receptor causing hypoglycaemia.

In this context, it is important to recall that the synthesis of IGF I under the control of GH never leads to hypoglycaemia or increased insulin sensitivity but rather to insulin resistance and impaired glucose tolerance as is typical for acromegaly. These effects on glucose metabolism are mostly due to increased secretion of GH leading to a stimulation of lipolysis and of gluconeogenesis while the sensitivity of muscle towards insulin is decreased.

IGFs and IGFBPs are Autocrine/Paracrine Cytokines

The somatomedin hypothesis of Salmon and Daughaday [24], according to which GH itself does not lead to growth but rather acts via a stimulation of IGF I synthesis in the liver, has been proven to be correct in many animal studies in which rhIGF I was administered. rhIGF I restores growth in hypophysectomized rats, has similar growth-promoting effects in mice which are genetically GH-deficient and is anabolic and growth-promoting in growth-arrested severely diabetic rats. During the last few years, rhIGF I has also been administered therapeutically over a period of several years to GH-insensitive (Laron) dwarfs who showed a very good growth response to rhIGF I. Therefore, one can conclude that IGF I can act as an endocrine hormone when released by the liver into the circulation and also when administered subcutaneously or intravenously to animals and humans.

However, this is not the whole story. IGF I and II are also produced by many other cells in the body, as are some of the binding proteins. In the embryo, the autocrine/paracrine effects of IGFs on differentiation of cells and tissues is a prerequisite for normal development. In tissue culture, IGFs and some of their binding proteins accumulate and are among the main so-called "conditioners" of culture media. It is believed that some of the binding proteins target IGFs to their receptors and/or keep IGFs from being degraded so that their activity is prolonged. In other instances, they prevent IGFs from reacting with their receptor. The impact of the presence of binding proteins on the activity of IGFs in different cell systems is still a major and much debated research topic.

Nevertheless, it is evident that IGFs are not only important endocrine hormones but also cytokines that stimulate cell differentiation and proliferation. Under the influence of IGF I, chondrocytes and osteoblasts stay differentiated and the synthesis of specific matrix proteins is stimulated. The same holds true for endocrine cells which increase their specific enzymatic equipment under the influence of IGF and thereby respond better to the specific trophic pituitary hormone. Furthermore, IGF I appears to have a protective effect against programmed cell death (apoptosis) [25]. Antiapoptotic effects of IGF I have been demonstrated in vitro as well as in vivo after cerebral and cardiac insults. In cardiac hypertrophy due to clamping of the aorta, the administration of rhIGF I improves cardiac function by stimulating the synthesis of contractile myofibrils and it increases the weight gain of rats [26].

Is There a Role for rhIGF I as a Therapeutic Tool in Diabetes Mellitus

IGF I levels tend to be decreased in diabetics. This is particularly evident in poorly controlled adolescent type 1 diabetics. These low levels of IGF I in labile type 1 diabetics correlate with increased GH secretory peaks during the night. These findings were the basis for clinical trials with rhIGF I. rhIGF I was administered in substitution amounts to poorly controlled type 1 diabetic adolescents in studies carried out by Dunger et al. [27]. It was found that with one subcutaneous injection of a relatively small dose of rhIGF I at bedtime, IGF I levels were normalized, excessive GH peaks during the night eliminated and diabetic control improved. The longest studies lasted 1 month and 3-month studies are now under way. The preliminary results of these trials suggest that IGF I may well be a useful adjuvant to insulin therapy for optimal metabolic control and somatic development of diabetic children.

In NIDDM, plasma IGF I levels are also dependent on the degree of metabolic control. In well controlled diabetics, IGF I levels are (near) normal, whereas they tend to be low in poorly controlled patients. Considering the relative insulin resistance, elevated lipid levels and compromised microcirculation in such patients, IGF I therapy may well improve control and decrease overall cardiovascular risk.

rhIGF I Effects in Normal Subjects

In large doses, rhIGF I is a hypoglycaemic substance. IGF I induces a short-lived hypoglycaemia in normal and in hypophysectomized rats. In man, the hypoglycaemic potency of rhIGF I is approximately 1/12 of that of insulin. When 100 μg of IGF I are administered per kilogram body weight as an intravenous bolus, the blood sugar falls very rapidly and to the same extent as in a classical insulin tolerance test with 0.15 U/kg body weight [28]. The recovery from hypoglycaemia occurs at about the same rate after either insulin or rhIGF I. When rhIGF I is administered subcutaneously or intravenously, total glucose consumption by the organism is increased. This increase of glucose oxidation and nonoxidative glucose disposal was quantitated during euglycaemic clamping with both hormones and again a potency ratio of 12:1 in favour of insulin was found [29]. The question arises whether rhIGF I acts mostly through the type 1 IGF or through the insulin receptor. This question can be answered in the case of bolus injections of rhIGF I and also when very large amounts of rhIGF I are infused. Under these circumstances, free fatty acid levels fall rapidly proving that IGF I in large concentrations also inhibits lipolysis in vivo as it does in vitro. Hepatic glucose release is also inhibited by large doses of rhIGF I [30]. Since neither tissue contains type 1 IGF receptors, this action of rhIGF I is bound to be mediated via the insulin receptor. Generally speaking, large concentrations of free IGF I in serum have acute insulin-like effects, mostly via cross-reaction with the insulin receptor. The controversial data in the literature regarding the effects of IGF I on hepatic glucose output and lipolysis can be explained by differences in dosing of rhIGF I. Very large doses always have an insulin-like effect also on the liver and on adipose tissue, whereas small doses of IGF I do not have such an effect. This statement is also supported by findings in patients with nonfunctional insulin receptors. Schoenle et al. [31] administered bolus injections of rhIGF I to three patients with type A insulin resistance. In all three patients, no acute hypoglycaemia occurred while there was a very slow fall of the blood sugar. The most likely explanation for the lack of a response to large doses of rhIGF I in patients with nonfunctional insulin receptors is that rhIGF I cannot cross-react with the defective insulin receptor and that the slow effects on glucose metabolism observed in these patients are mediated via the type 1 IGF receptor. The results of these experiments also support the notion that the signal transduction through the type 1 IGF receptor is different from that through the insulin receptor. In any case, the Glut-4 response which is so closely connected to the binding of insulin to its receptor does not appear to be shared by the interaction between IGF I and the type 1 receptor. It is important to recall the results of experiments using high doses of rhIGF I in streptozotocin-diabetic rats. These rats responded to rhIGF I with a growth response in spite of the fact that hyperglycaemia, glucosuria and overall glucose metabolism were not improved by rhIGF I treatment. These data again show that even though rhIGF I may cross-react with the insulin receptor, this is not a major component of the organism's response to IGF I.

General Observations and Reasons in Favour of rhIGF I

Why should rhIGF I ever play a part in the therapy of diabetes mellitus if it cannot be administered in doses in which it directly acts on glucose uptake, thereby lowering blood glucose? In the following we could like to discuss several observations which make us believe that rhIGF I may eventually play a role in the therapy of different forms of diabetes mellitus (Fig. 5):

1. As a consequence of the metabolic disturbances of diabetes, i.e. of the ineffectiveness of insulin, GH-hypersecretion and poor utilization of nutrients, the synthesis and secretion of IGF I from the liver is decreased. This has only little to do with the insulin dose that is administered to diabetics, but rather with the metabolic disturbance resulting from hypoglycaemia alternating with hyperglycaemia and the poor response of the "diabetic" liver to insulin. In very well controlled type 1 diabetics, serum IGF I levels are normal and insulin sensitivity of tissues is also normal. However, with the present schemes of insulin therapy, it is practically impossible to achieve ideal metabolic control so that most type 1 diabetics do have a certain degree of IGF I deficiency. The same holds true for type 2 diabetics who could theoretically be in optimal metabolic control if they lost weight on a reducing regimen. However, this appears to be particularly difficult for many type 2 diabet-

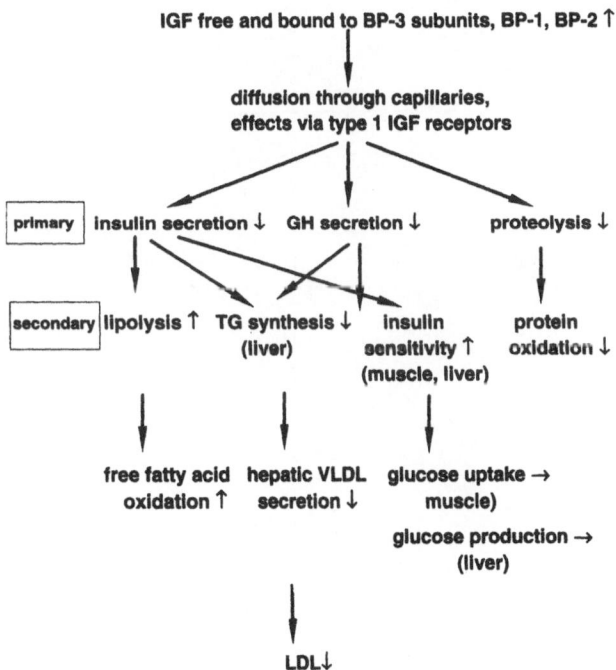

Metabolic and endocrine sequelae of IGF I therapy

Fig. 5. Metabolic and endocrine sequelae of insulin-like growth factor (IGF-I) therapy. *BP*, binding protein; *VLDL*, very low-density lipoprotein; *LDL*, low-density lipoprotein. From [35]

ics who may have inherited a set of "insulin resistance" genes with a tendency to hyperphagia.

2. Uncontrolled type 1 diabetics and obese type 2 diabetics are relatively insulin-resistant. Insulin resistance is aggravated in NIDDM by obesity. All diabetics characteristically have elevated levels of insulin in the peripheral blood. Thus, type 1 diabetics treated with one or two daily injections of insulin have for the most part

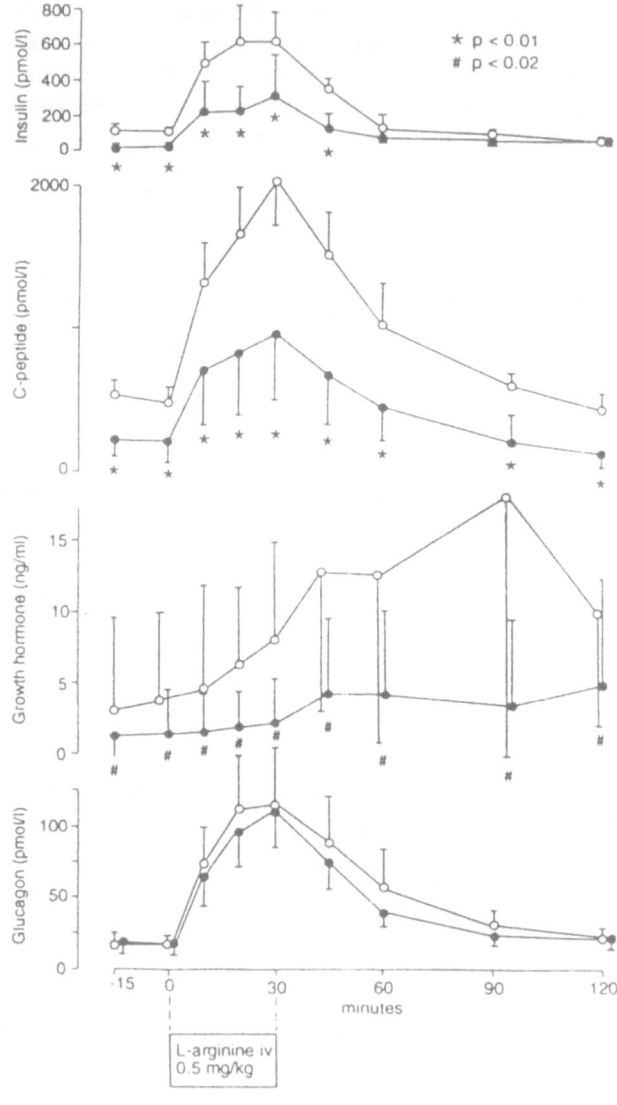

Fig. 6. Venous insulin, C-peptide growth hormone and glucagon during L-arginine stimulation on day 3 of saline (*open circles*) or rhIGF-I (10 μg/kg per hour s.c., *closed circles*) treatment in eight healthy subjects (mean ± SD). From [34]

of the day higher peripheral insulin levels than normal subjects. However, insulin peaks are lower but last longer. The same is true for patients with NIDDM who have increased endogenous insulin levels while fasting, during the night and in the morning but respond slowly to an increase of glycaemia after meals and characteristically exhibit delayed insulin peaks. Despite elevated insulin levels, most of these patients are hyperglycaemic during most of the day, and hyperglycaemia aggravates the sluggishness of the response of the β-cells to hyperglycaemia and other stimuli (Fig. 6).

3. rhIGF I administered intravenously or subcutaneously has two major endocrine effects: (a) it partially inhibits insulin secretion [32,33], and (b) it decreases the number and the height of GH peaks during the night and decreases GH secretion after secretory stimuli [34]. One might expect that the partial inhibition of insulin secretion might result in hyperglycaemia. However, the reverse is the case. In normal subjects as well as in early type 2 diabetics, the administration of rhIGF I leads to an improvement of glucose tolerance at comparatively lower insulin secretion rates [35]. Only when insulin secretion is further inhibited by much larger doses of rhIGF I resulting in fasting hypoglycaemia, glucose intolerance may result.

4. In order to determine whether rhIGF I therapy acts mostly via suppression of GH secretion which, when excessive, is known to induce hyperinsulinaemia and glucose intolerance, rhIGF I was administered to GH-deficient adults. In GH-deficient patients, rhIGF I therapy also improved glucose tolerance at decreased insulin levels indicating that inhibition of GH secretion is not the sole or major reason why rhIGF I therapy improves glucose tolerance and increases the organism's sensitivity to insulin [36] (Fig. 7).

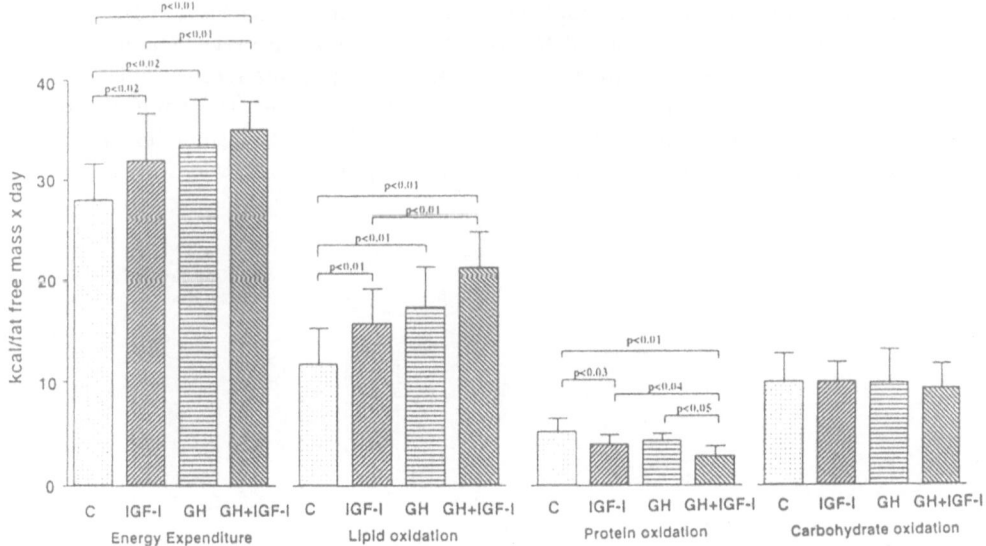

Fig. 7. Resting energy expenditure and lipid, protein, and carbohydrate oxidation rates in eight growth hormone-(GH) deficient subjects on day 7 of indicated treatment (mean ± SD). *IGF*, insulin-like growth factor; *C*, control. From [36]

5. In both normal and GH-deficient subjects, the administration of recombinant IGF I in low doses leads to a rise of fasting and postprandial free fatty acid concentrations because of the decreased insulin levels. By loosening the insulin brakes on lipolysis, free fatty acid levels and ketone body levels are increased by rhIGF I therapy. According to the classical glucose fatty acid cycle of Randle, an increase of free fatty acids and ketone bodies should lead to decreased glucose utilization and loss of insulin sensitivity. However, this is not the case during the administration of rhIGF I. In normal subjects as well as in GH-deficient subjects, the following observations were made: rhIGF I increased total energy expenditure of the organism at the expense of increased lipid oxidation, while glucose oxidation and nonoxidative glucose utilization remained constant and protein oxidation was reduced at decreased amino acid levels [34,36]. Therefore, the administration of rhIGF I switches the organism to lipid oxidation at unchanged glucose utilization rates and increases insulin sensitivity. Increased insulin sensitivity was actually demonstrated during euglycaemic, hyperinsulinaemic clamping during constant rhIGF I administration. Glucose utilization was clearly increased during rhIGF I administration at comparable insulin concentrations.

6. Dunger et al. have used rhIGF I as a substitution therapy giving one dose of rhIGF I at night to adolescent type 1 diabetics. IGF I levels were raised from below normal to normal levels, glycaemic control was improved at insulin doses which were decreased by 10%–20%. Glycosylated haemoglobin levels were significantly improved after 1 month of IGF I therapy and no side-effects were noted at these low doses of rhIGF I. These preliminary results of rhIGF I substitution therapy in adolescent type 1 diabetics make it likely that rhIGF I therapy as an adjuvant to insulin therapy may become a very useful tool in rendering glycaemic control easier for type 1 diabetics while improving the prognosis with respect to diabetic late complications. rhIGF I therapy has been experimentally tried in a small number of NIDDM patients who had previously been on dietary control plus or minus sulfonyl ureas. These patients received relatively larger doses of rhIGF I and responded very well with respect to significant improvement of glycaemic control at decreased endogenous insulin levels and an improvement of the lipid profile. Within 5 days of fasting, very low-density

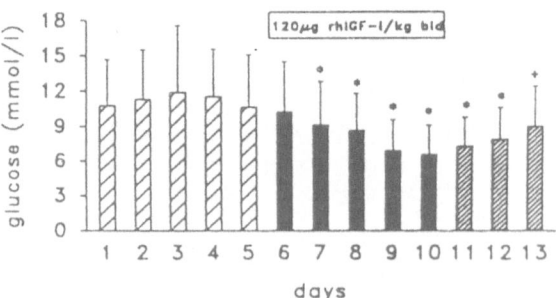

Fig. 8. Fasting plasma glucose in type 2 diabetics before, during and after s.c. rh insulin-like growth factor 1. $^\dagger p < 0.02$ v. control; $^* p < 0.01$ v. control. From [35]

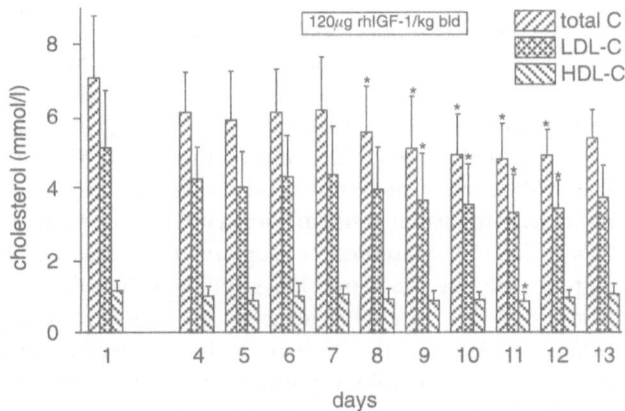

Fig. 9. Total cholesterol, low-density lipoprotein (LDL)- and high-density lipoprotein (HDL)-cholesterol in type 2 diabetes before, during and after subcutaneous rh insulin-like growth factor (IGF) I. From [37]

lipoprotein (VLDL)-triglyceride levels were significantly improved and low-density lipoprotein (LDL)-cholesterol levels decreased (Figs. 8, 9).

Cardiovascular Effects of rhIGF I

In the first experiments with rhIGF I in normal subjects, an increased renal plasma flow and glomerular filtration rate was observed [38]. Later it was shown that during rhIGF I administration, blood flow through skeletal muscle is increased. Vasodilatory effect in the kidney and in muscle is dependent on the activation of the NO synthase since the effect of IGF I can be blocked by the addition of NO synthase inhibitors [39]. The effect of rhIGF I on vasodilation in other areas of the body has not yet been studied. However, it is possible that some of the side-effects of large doses of rhIGF I may be related to the vasodilatory effect of IGF I: headaches, cerebral palsy, oedema in the lower extremities and in the face and possibly also instances in which skeletal pain was located to muscles which may have been oedematous and swollen.

On the other hand, the vasodilatory effect of rhIGF I may be very useful in patients with coronary artery disease and arteriosclerotic lesions of the lower limbs and it may also be one factor responsible for increased glucose uptake by muscle.

The Role of Albert Renold in the IGF Story

After graduating from Medical School in Zurich, I had the good luck to be accepted as a research fellow in Dr. Thorn's laboratory at the Peter Bent Brigham Hospital which, at that time, was run by Albert Renold. There I started my research and academic career, encouraged and always supported by Albert Renold. The fat pad assay for determination of insulin-like activity was then one of the main research projects in Albert Renold's laboratory and this is also where I found out what epidydimal adipose tissue might be good for. His enthusiasm and open mind for new ideas and his continuous encouragement when things in the laboratory did not go so well were essential for the discovery of the IGFs. When the scientific world decided that NSILA was a non-entity and should not be called insulin-like because it did not react with insulin antibodies and could not be measure with a radioimmunoassay, Albert Renold continued to support our research, to exchange ideas and to help in solving problems. He strongly believed in biology, physiology and in progress if the underlying ideas were sound and if hypotheses were worth being tested. I wish to thank him here for his intellectual input into the discovery of the impact of IGF I in biology, physiology and pathophysiology of growth and diabetes and for his friendship and empathy which gave me the courage to "survive" in diabetes research.

Conclusions

The term NSILA was coined in 1960 and the two peptides responsible for non-suppressible insulin-like activity were isolated, characterized and sequenced by Rinderknecht and Humbel in 1976. From then on IGF research exploded and it became possible to solve many questions regarding the synthesis, secretion, regula-tion, biological effects and physiological importance of the IGFs. IGF I has been found to be the most important mediator of the action of pituitary GH. The knowledge about this function of rhIGF I has resulted in the very successful treatment of GH-insensitive Laron dwarfs with IGF I. Whereas GH-deficient dwarfs grow well if treated with pituitary GH, GH-insensitive Laron dwarfs grow similarly well when treated with rhIGF I.

Whereas glucose homeostasis is fully dependent on the second-to-second regula-tion of insulin secretion by the β-cell, IGF I may be involved in the overall sensitivity of the organism to insulin. When nutrition is inadequate, IGF I levels fall and the organism develops a certain degree of insulin resistance. IGF I plays a major role in increasing insulin sensitivity again during refeeding and in the fed state. The balance between insulin sensitivity and insulin resistance is influenced by IGF I. This can be shown very convincingly when relatively insulin-resistant type 1, type 2 or type A diabetics are treated with rhIGF I. During treatment with rhIGF I, the organism becomes more sensitive to insulin and glucose utilization occurs at considerably lower insulin levels. A relative IGF I deficiency may be related to the so-called insulin resistance genes which are held responsible for NIDDM explosion in ethnic groups which had lived mostly on protein, not having developed agriculture, and then were suddenly transferred into the Western world with a nutritive surplus of carbohydrate and fat.

The physiology and pathophysiology of IGF I have become particularly complex when it was found that IGF I is not only an endocrine hormone but also a very important autocrine/paracrine cytokine functioning, for instance, as an estramedin in the process of remodelling of bone. Even more difficult is the interpretation of the physiological and pathophysiological roles of the six different, specific IGFBPs. These latter aspects of the physiology of IGF are now being investigated in a large number of laboratories all over the world.

Although many questions remain to be answered, we believe that rhIGF I will find a place as an important player in diabetes therapy. It increases insulin sensitivity, helps burn fat, lowers LDL-cholesterol levels and has a vasodilatory effect so that it may well become an interesting adjuvant to insulin.

Acknowledgements. The work on which this review is based was always supported by the Swiss National Science Foundation (present grant No 32-31281.91). Ciba made rhIGF I available and sponsored some of the basic research and the clinical protocols with rhIGF I. Jürgen Zapf was the major contributor to most original studies. My special thanks go also to Christoph Schmid and to the whole IGF team. I also wish to thank Mrs. Martha Salman for maintaing very high standards in the secretariat and for her expert help in preparing this manuscript. Our late research director of medicine Alexis Labhart made it possible for our team to work under favourable circumstances and in a climate of mutual respect and friendship. He was also a very good and faithful friend of Albert E. Renold.

References

 1. Rinderknecht E, Humbel RE (1978) The amino acid sequence of human insulin-like growth factor I and its structural homology with proinsulin. J Biol Chem 253: 2769
 2. Humbel RE (1990) Insulin-like growth factors I and II. Eur J Biochem 190: 445
 3. Froesch ER, Zapf J (1985) Insulin-like growth factors and insulin: comparative aspects. Diabetologia 28: 485–493
 4. Ullrich A, Gray A, Tam A et al (1986) Insulin-like growth factor I receptor primary structure: comparison with insulin suggests structural determinants that define functional specificity. EMBO J 5: 2503
 5. De Meyts P. Personal communication
 6. Morell B, Froesch ER (1973) Fibroblasts as an experimental tool in metabolic and hormone studies. II. Effects of insulin and nonsuppressible insulin-like activity (NSILA–S) on fibroblasts in culture. Eur J Clin Invest 3: 119–123
 7. Schmid C, Steiner T, Froesch ER (1984) Insulin-like growth factor I supports differentiation of cultured osteoblast–like cells. FEBS Lett 173: 48–52
 8. Ernst M, Froesch ER (1987) Osteoblastlike cells in a serum-free methylcellulose medium form colonies: effects of insulin and insulinlike growth factor I. Calcif Tissue Int 40: 27–34
 9. Laron Z, Anin S, Klipper-Aurbach Y et al (1992) Effects of insulin-like growth factor on linear growth, head circumference, and body fat in patients with Laron-type dwarfism. Lancet 339: 1258
10. Walker JL, van Wyk JJ, Underwood LE (1992) Stimulation of statural growth by recombinant insulin-like growth factor I in a child with growth hormone insensitivity syndrome (Laron-type). Pediatrics 121: 641
11. Scheiwiller E, Guler H-P, Merryweather J, Scandella C, Maerki W, Zapf J, Froesch ER (1986) Growth restoration of insulin-deficient diabetic rats by recombinant human insulin-like growth factor I. Nature 323: 169–171
12. Efstratiadis A (1994) IGFs and dwarf mice: genetic and epigenetic control of embryonic growth. In: Sizonenko PC, Aubert ML, Vassalli J-D (eds) Developmental endocrinology, vol 6. Ares-Serono Symposia Series. Raven, New York, p 27 (Frontiers in endocrinology)
13. Rabinowitz DK, Zierler L (1963) A metabolic regulation device based on the actions of human growth hormone and of insulin, singly and together, on the human forearm. Nature 199: 913–914
14. Froesch ER, Bürgi H, Ramseier EB, Bally P, Labhart A (1963) Antibody-suppressible and nonsuppressible insulin-like activities in human serum and their physiologic significance. An insulin assay with adipose tissue of increased precision and specificity. J Clin Invest 42: 1816–1834

15. Brand Miller JC, Colagiuri S (1994) The carnivore connection: dietary carbohydrate in the evolution of NIDDM. Diabetologia 37: 1280–1286
16. Neel JV (1962) Diabetes mellitus: a "thrifty" genotype rendered detrimental by "progress"? Am J Hum Genet 14: 353–362
17. Kaufmann U, Zapf J, Torretti B, Froesch ER (1977) Demonstration of a specific serum carrier protein of nonsuppressible insulin-like activity in vivo. J Clin Endocrinol Metab 44: 160–165
18. Zapf J, Jagars G, Sand I, Froesch ER (1978) Evidence for the existence in human serum of large molecular weight nonsuppressible insulin–like activity (NSILA) different from the small molecular weight forms. FEBS Lett 90: 135–140
19. Guler H-P, Zapf J, Schmid C, Froesch ER (1989) Insulin-like growth factors I and II in healthy man. Estimations of half-lives and production rates. Acta Endocrinol (Kbh) 121: 753–758
20. Baxter RC, Martin JL (1989) Structure of the Mr 140 000 growth hormone-dependent insulin-like growth factor binding protein complex: determination by reconstitution and affinity-labelling. Proc Natl Acad Sci USA 86: 6898
21. Zapf J, Hauri C, Waldvogel M, Froesch ER (1986) Acute metabolic effects and half-lives of intravenously administered insulin–like growth factors I and II in normal and hypophysectomized rats. J Clin Invest 77: 1768–1775
22. Zapf J, Futo E, Peter M, Froesch ER (1992) Can "big" IGF II in serum of tumor patients account for the development of extrapancreatic tumor hypoglycemia (EPTH)? J Clin Invest 90: 2574–2584
23. Daughaday WH, Kapadia M (1989) Significance of abnormal serum binding of insulin-like growth factor II in the development of hypoglycemia in patients with non-islet-cell-tumors. Proc Natl Acad Sci USA 86: 6778–6782
24. Salmon WD Jr, Daughaday WH (1957) A hormonally controlled serum factor which stimulates sulfate incorporation by cartilage in vivo. J Lab Clin Med 49: 825–836
25. Gluckmann PD, Beilharz EJ, Johnston BM et al (1994) Growth factors and perinatal asphyxia. In: Sizonenko PC, Aubert ML, Vassalli J-D (eds) Developmental endocrinology, vol 6. Ares-Serono Symposia Series. Raven, New York, p 185 (Frontiers in endocrinology)
26. Donath MY, Zapf J, Eppenberger-Eberhardt M, Froesch ER, Eppenberger HM (1994) Insulin-like growth factor I stimulates myofibril development and decreases smooth muscle alpha-actin of adult cardiomyocytes. Proc Natl Acad Sci USA 91: 1686–1690
27. Dunger DB (1994) IGF-I treatment of type I diabetic subjects. In: Annual Meeting of the European Society of Clinical Investigation. 20–23 April 1994, Toledo, Spain
28. Guler HP, Zapf J, Froesch ER (1987) Short-term metabolic effects of recombinant human insulin-like growth factor I in healthy adults. N Engl J Med 317: 137–140
29. Turkalj I, Keller U, Ninnis R et al (1992) Effect of increasing doses of recombinant human insulin-like growth factor-I on glucose, lipid, and leucine metabolism in man. J Clin Endocrinol Metab 75: 1186–1191
30. Jacob RE, Barrett E, Plewe G et al (1989) Acute effects of insulin–like growth factor I on glucose and amino acid metabolism in the awake fasted rat. J Clin Invest 83: 1717–1723
31. Schoenle EJ, Zenobi PD, Torresani T, Werder EA, Zachmann M, Froesch ER (1991) Recombinant human insulin-like growth factor I (rhIGF 1) reduces hyperglycaemia in patients with extreme insulin resistance. Diabetologia 34: 675–679
32. Guler HP, Schmid C, Zapf J, Froesch ER (1989) Effects of recombinant insulin-like growth factor I on insulin secretion and renal function in normal human subjects. Proc Natl Acad Sci USA 86: 2868–2872
33. Zenobi PD, Graf S, Ursprung H, Froesch ER (1992) Effects of insulin-like growth factor-I on glucose tolerance, insulin levels, and insulin secretion. J Clin Invest 89: 1908–1913
34. Hussain MA, Schmitz O, Mengel A, Keller A, Christiansen JS, Zapf J, Froesch ER (1993) Insulin-like growth factor I stimulates lipid oxidation, reduces protein oxidation, and enhances insulin sensitivity in man. J Clin Invest 92: 2249–2256
35. Zenobi PD, Jaeggi-Grosiman SE, Riesen W, Roder M, Froesch ER (1992) Insulin-like growth factor-I improves glucose and lipid metabolism in type 2 diabetes mellitus. J Clin Invest 90: 2234–2241
36. Hussain MA, Schmitz O, Mengel A, Glatz Y, Christiansen JS, Zapf J, Froesch ER (1994) Comparison of the effects of growth hormone and insulin-like growth factor I on substrate oxidation and on insulin sensitivity in growth hormone-deficient humans. J Clin Invest 94: 1126–1133

37. Zenobi PD, Holzmann P, Glatz Y, Riesen WF, Froesch ER (1993) Improvement of lipid profile in type 2 (non-insulin-dependent) diabetes mellitus by insulin-like growth factor I. Diabetologia 36: 465–469
38. Guler H-P, Eckardt K-U, Zapf J, Bauer, Froesch ER (1989) Insulin-like growth factor I increases glomerular filtration rate and renal plasma flow in man. Acta Endocrinol (Kbh) 121: 101–106
39. Hussain MA, Schmitz O, Christiansen JS, Zapf J, Froesch ER (1994) Insulin-like growth factor-I increases forearm blood flow at a non-hypoglycemic dose. (Abstract). 15th International Diabetes Federation Meeting. Nov 1994, Kobe, Japan

On the Pathogenesis of Insulin-Dependent Diabetes Mellitus in Man: A Paradigm in Transition

J. Nerup, T. Mandrup-Poulsen, F. Pociot, A.E. Karlsen, H.U. Andersen, U.B. Christensen, T. Sparre, J. Johannesen, and O.P. Kristensen

In any epoque and in any field of biomedicine a few major figures stand out as being particularly influential. Some because they contributed very significantly to science, some because their personalities made them natural centres of learning and research. Albert E. Renold did both and will be remembered as one of the greatest names in diabetology of this century. His curiosity, his emphatic interest in young people and their ideas, his open-mindedness and intuitive understanding of new and sometimes rather vaguely formulated concepts made conversations with Albert Renold delightful and encouraging experiences. One of us (J.N.) remembers a long conversation in Geneva several years ago where seemingly odd and certainly less than solidly substantiated ideas on the pathogenesis of insulin-dependent diabetes mellitus (IDDM) were discussed. Pointing out these shortcomings Renold said, quoting Thomas Huxley: "He who does not go beyond the facts, he will seldom get as far as the facts". With this in mind, however, still going to the limits of the facts and probably beyond, we wish to dedicate this chapter on the pathogenesis of IDDM to the memory of Albert E. Renold.

Insulin-Dependent Diabetes Mellitus is an Autoimmune Disease: Establishing a Paradigm

Insulitis, the pathoanatomical hallmark of IDDM, was first reported around the turn of the century [1], but attracted only sporadic interest for years until its rediscovery by Gepts in 1965 [2]. In the 1950s and 1960s humoral thyroid and adrenal autoimmunity were described [3,4], and in the late 1960s cell-mediated autoimmunity was demonstrated in these conditions [5] by means of rather primitive methodology. In vitro and in vivo demonstration of cell-mediated immunoreactivity to crude preparations of endocrine pancreatic tissue was accomplished in 1971 [6]. Shortly thereafter followed the discovery of islet cell antibodies (ICA) [7], the human leucocyte antigen (HLA)-IDDM association with alleles of HLA-class I [8], and subsequently of the HLA-class II region [9]. Simultaneously, the possible role of virus in the aetiology of IDDM was much discussed [10]. These findings prompted us to propose a comprehensive hypothesis concerning the aetiology and pathogenesis of IDDM: "One or more immune-response genes associated with HL-A8 and/or W15 might be responsible for an altered T-lymphocyte response. The genetically determined host response could fail to eliminate an infecting virus (coxsackie virus B4 and others) which in turn might destroy the pancreatic beta-cells or trigger an autoimmune reaction against the infected organ" [8]. On the basis of the flow of new information this hypothesis was modified in 1981 to: "Genetic predisposition (susceptibility) to IDDM is conferred by two genes on chromosome 6, one associated with HLA-D/DR3, one with HLA-D/DR4. The reaction of susceptible individuals to certain environmental stimuli (beta-cell cytotoxic virus? beta-cell cytotoxic chemical?) is abnormal, leading to: beta-cell destruction directly, through "autoimmune mechanisms" or because of lack of regeneration of the beta-cell after damage."

Interest in the field grew rapidly and with that the literature. To date more than 7000 papers on the immunology and immunogenetics of IDDM have been published. The early findings were confirmed, extended and detailed. The 64-kDa autoantibody, the antigens of which were later found to be glutamic acid decarboxylases (GADs) [11], was identified [12]. Focus on the HLA-IDDM association shifted from DR3 and DR4 to specific alleles of the DQB1 and DQA1 loci (see below) [13]. Sophisticated studies, in particular spleen-cell transfer experiments and transfer of islet-reactive T-cell clones in the spontaneous animal models of IDDM provided detailed information on the roles played by the T-lymphocyte subsets, i.e., CD4+ and CD8+ T-lymphocytes in the insulitis process and thereby, presumably, in the IDDM pathogenesis in these animals [14,15]. It was further demonstrated that cytokines were cytotoxic to beta-cells [16,17] and it was suggested that the balance between Th1 lymphocytes characterized by a cytokine secretory profile dominated by interferon-γ (IFN-γ) and interleukin (IL)-2, i.e., a cytotoxic response, and Th2 lymphocytes, characterized by an IL-4, IL-10 and transforming growth factor β (TGFβ) secretory profile, i.e., a non-aggressive antibody-dominated immune response [18], might be biased towards the Th1 type [19]. The hypothesis that IDDM is a classical autoimmune disease rapidly gained wide acceptance and was soon established as an almost unchallengable paradigm. In short, the implications of these observations would be:

- Genetic susceptibility to IDDM is mainly conferred by HLA-DQ genes.
- The Th1–Th2 lymphocyte balance is biased towards Th1.
- Beta-cells are destroyed by cytotoxic T-lymphocytes (CD4+ and/or CD8+) characterized by a Th1 cytokine secretory profile.
- Beta-cells present autoantigen on their cell surfaces to be recognized by T-lymphocytes through receptors [T-cell receptors (TCR)].
- This recognition may be initiated by environmental antigens sharing antigenic peptide sequences with beta-cell proteins (molecular mimicry).
- GAD65 is the primary beta-cell antigen.

In other words, the central problem in the pathogenesis of IDDM was seen as a problem of the immune system.

Is Insulin-Dependent Diabetes Mellitus a Classical Autoimmune Disease? A Paradigm Under Pressure

Ample evidence exists to support the premise that the classical immune mechanisms are involved in beta-cell destruction [15]. However, meticulous time sequence studies in the BB-rat [20] and the NOD mouse [21] of the building-up of the cellular infiltrate in the islets, i.e., insulitis, have shown that cells of the monocyte–macrophage–dendritic cell lineage (Mo/Mø/DC) are the first to appear, and that CD4+, CD8+ and B-lymphocytes follow in this order of sequence. Prior to this, focal beta-cell necrosis can be seen. We had shown [22] that supernatants of mitogen stimulated non-diabetic human peripheral blood monocytes from healthy men were cytotoxic to rat and human beta-cells in vitro, and subsequently identified IL-1, potentiated by tumour necrosis factor α (TNFα) as the cytotoxic principle [16,17]. IL-1 and TNF are secretory products of activated Mo and Mø (and DC?) and in the islets such cells were shown to contain abundant mRNA for IL-1 and TNFα [23,24].

Ablation experiments with silica [25,26] showed that functional macrophages are obligatory for the disease process, whereas passive transfer experiments of spleen cells and specifically reactive T-cell clones to young NOD mice and young NOD/SCID mice suggested that CD8+ cytotoxic T-lymphocytes are not necessary and CD4+ lymphocytes not sufficient to produce IDDM in the recipients [14].

Hence, the earliest events affecting the target beta-cell, i.e., the pathogenesis *in sensu strictu*, may not require the involvement of the T-lymphocyte system. The Mø/Mo/DCs play key roles as producers of beta-cell cytotoxic cytokines, as initiators and primary accelerators of the beta-cell cytotoxic immune response. This strongly suggests that beta-cell destruction in IDDM has two distinct phases:

1. A non-antigen driven, non-lymphocyte mediated initiation phase
2. An antigen driven (multiple antigens?), lymphocyte mediated amplification and perpetuation phase.

Conventional thinking and the strong HLA-IDDM association suggest the existence of a primary antigen initiating the destructive immune response responsible for beta-cell destruction in individuals genetically susceptible to development of IDDM. In the NOD mouse there is evidence [27], however, yet to be convincingly confirmed, that GAD fulfils this role. The notion that GAD might be the initiating primary antigen also in human IDDM seems to be shared by many investigators [28].

It is difficult to accept GAD either as the initiating, primary or even major autoantigen driving the beta-cell destructive cellular immune response in IDDM. Accordingly, GADs are not beta-cell specific enzymes (GADs are present in brain, nerves, islet alpha-cells and other tissues) and GAD "autoantibodies" are almost equally frequent in newly diagnosed IDDM patients with or without IDDM associated HLA-class II alleles and genotypes [29]. As can be seen from Table 1, similar arguments can be made concerning ICA, insulin autoantibodies (IAA), 155 KD and ICA 512, and probably all the other "autoantibodies".

Table 1. The prevalence of putative autoantibodies in relation to HLA-DR phenotype: lack of HLA-DR restriction

HLA-DR Type	ICA (%)	IAA(%)	GAD65 (%)	155K (%)
3–x	41	21	64	87
3–4	56	30	64	88
4–x	43	23	70	89
x–x	31	16	73	88

HLA, human leucocyte antigen; ICA, islet cell antibodies; IAA, insulin autoantibodies; GAD, glutamic acid decarboxylase.

Table 2. Putative Autoantigens in IDDM

Insulin–proinsulin
GAD65
GAD67
ICA 512/IA2 (tyrosine phosphatase)
Carboxypeptidase H
Gangliosides
ICA69
Peripherin (58 kDa)
Heat shock proteins 62 and 65
Glut 2
155 kDa rat insulinoma cell membrane protein
p52 rat insulinoma cell membrane protein
32, 55–72, 120–170 kDa islet proteins
Sulfatides

GAD, glutamic acid decarboxylase; ICA, islet cell antibody.

Insulin is, in fact, the only beta-cell specific "autoantigen" identified so far. But even insulin autoantibody formation is not HLA-Class II restricted and only 35%–40% of newly diagnosed IDDM patients present with these autoantibodies. Furthermore, native insulin has been secreted to the circulation since the 16th–17th week of gestation in humans, and is therefore available for "self-tolerization" in foetal life. Even the mere number of putative autoantigens (Table 2) is difficult to reconcile with the existence of an initiating, primary and specific T-lymphocyte-dependent beta-cell destructive insult driven by one antigen.

Available information about T-cell reactivity to these antigens is sparse. However, no evidence suggesting HLA-restriction of T-cell responses to 38K insulin secretory granule membrane protein (imogen 38), ICA 69 and GAD65 exists (personal communications, B. Roep, Roep and Kolb, A. Worsaa). The animal models of IDDM, the NOD mouse in particular, facilitate studies of T-cell reactivity to IDDM-associated autoantigens to some extent. In the NOD mouse, GAD autoimmunity develops early and reactivity to carboxypeptidase H, HSP65 and insulin subsequently develop with time, closer to the onset of disease [27]. This may suggest that GAD is in fact the primary autoantigen in the NOD mouse, but no data support that GAD serves this role in the BB-rat [30]. Furthermore, experimentally induced strong T-cell reactivity against GAD65 in non-diabetic strains [31] and transfer of CD4 T-cell clones reactive against GAD did not produce IDDM in disease-free recipients. Hence, even in the NOD mouse the primary initiating autoantigen may still await identification.

T-cell reactivity assays are notoriously difficult to perform in a reproducible, quantitative fashion. They are important and should be developed since a Th1-type T-cell response in vitro and in vivo to relevant islet antigens is probably a more clinically significant parameter to identify an ongoing autoaggressive immune destruction of beta-cells than the demonstration of autoantibodies to such antigens. At present we do not know if a beta-cell specific initiating primary autoantigen responsible for IDDM in man does exist, and, if it does exist, it has yet to be identified.

Considering this, we hypothesized [32] that islet proteins are not antigenic in their native forms; rather, they become antigenic when and because beta-cells are destroyed through a common mechanism resulting from beta-cell destruction during the initiation phase (see below).

The literature contains an enormous number of papers analysing the HLA-IDDM association. The HLA-class II gene locus showing the strongest IDDM association is the DQ-locus and transracial studies have shown that relatively few DQB1 and DQA1 alleles are involved in genetic susceptibility to IDDM [33]. However, the differences in population allelic distribution of the IDDM-associated DQ specificities cannot alone explain the massive variations in IDDM incidence reported. The IDDM-associated alleles and genotypes are rather frequent in the general population. DQB1 non-Asp/non-Asp homozygocity is seen in approximately 20% of Scandinavians. In all Caucasian populations 5%–10% of young-onset IDDM patients, phenotypically indistinguishable from IDDM in general, cannot form an HLA-class II IDDM susceptibility heterodimer molecule on their immune cell surfaces [33]. These observations clearly indicate that the HLA-class II susceptibility genes are neither necessary nor sufficient to produce IDDM in man.

In addition, in some European Caucasian populations HLA-DQ is not the locus conferring the strongest statistical risk of IDDM. We recently showed that HLA-DRB1 molecules with lysine in position 71, i.e. DRB1*0401 were the strongest single alleles in association with IDDM. Seventy-four percent of patients and 15% of controls carry DRB1 Lys71[+], and linkage between IDDM and DRB1 Lys71[+] is strong in IDDM families [number of sibpair families 81, logarithm of differences (lod) score 4.26, $p < 10^{-6}$] and DRB1 Lys71[+/+] homozygous individuals have a relative risk of IDDM of 103.5. Further analyses of haplotype relative risks (HRR) revealed a much stronger effect of DRB Lys71[+] (HRR 8.38) than DQB1 non-Asp (HRR 2.53) [34].

Conventionally, the role of HLA-class II molecules is believed to be the presentation of endogenous antigenic peptides to the T-helper lymphocytes. The lack of beta-cell specificity, lack of specificity of HLA-association (restriction) and the large number of putative autoantibodies, together with the lack of specificity of the T-cell responses and the fact that the HLA-class II and disease association is not obligatory, make it difficult to maintain the view that IDDM is solely a problem of the immune system. There is little, if any, evidence to demonstrate that cytotoxic T-lymphocytes destroy beta-cells directly by cell-to-cell contact. Rather, it seems that cells of the immune system destroy beta-cells by means of secretory products (see below). Since there is no reason to believe that beta-cells alone are exposed to these immune mediators, it seems logical to suggest that the basic problem of IDDM is in the beta-cell as such.

The Pathogenesis of Insulin-Dependent Diabetes Mellitus: Some Elements of a New Paradigm

Even though the immunological and genetic phenomena involved in IDDM pathogenesis may lack specificity, the fact that beta-cells are specifically destroyed in IDDM needs to be explained by any model of IDDM pathogenesis. Our observation [22] that supernatants of mitogen-activated peripheral blood mononuclear cells (PBMC) could inhibit glucose-induced insulin secretion and specifically destroy beta-cells in isolated rat and human islets in culture strongly suggested that immune mediator molecules produced by PBMC might be operating. Subsequent studies by our group, later to be confirmed by others, identified IL-1β (IL-1) as the beta-cell cytotoxic substance [16], and showed that IL-1's effects were potentiated by TNFα (TNF) and IFN-γ (IFN) [17]. Morphological and functional data demonstrated the cytotoxic effect to be beta-cell specific. Since it was known that cytokines induce oxygenderived free radical (FOR) formation in certain cell lines and that beta-cells were known to have low Mn superoxide dismutase (SOD) (a free oxygen radical, FOR, scavenger) and catalase concentrations [35], we suggested [32] that cytokine induced FOR production in beta-cells not properly scavenged was the mechanism initiating beta-cell destruction. Direct experimental evidence to support this concept, i.e., demonstration of FOR in beta-cells, is still lacking. Good indirect evidence, however, is plentiful, e.g., effects of FOR scavengers [36] and upregulation of oxidative stress proteins by cytokines in a pattern only matched by islet exposure to H_2O_2 [37].

Chemically produced FOR and nitric oxide (NO), another free radical species, both destroy beta- and alpha-cells of isolated islets in vitro [38]. Since the enzyme producing NO, the cytokine inducible nitric oxide synthetase (iNOS), is induced only in beta-cells [39], these observations suggest that cytokine-induced intracellular NO and FOR formation in the beta-cells are mechanisms of major importance in the early phases of beta-cell destruction. Macrophages and endothelial cells in the islet may contribute by FOR and NO formation, although probably to a minor degree. Both free radical species may induce DNA strand breaks [40], and probably apoptotic cell death [41]. NO causes nitrosylation and inactivation of haeme-containing enzymes, notably respiratory reductases e.g. aconitase, leading to decreased ATP formation, and both FOR and NO may modify the structure of several proteins. We have suggested [32] that this may expose novel immunogenic linear epitopes on many beta-cell proteins and hence explain the multitude of putative autoantigens in IDDM. The immune system might see these as "non-self".

It follows that in beta-cells, as in any cell type carrying cytokine receptors, cytokines will induce a race between deleterious and protective (upregulation of stress proteins [37] and scavengers [42]) events and that in beta-cells the deleterious prevail. One possible way to study this in more detail would be to look at cytokine-induced changes in islet protein expression and post-translatory protein modification. We attempted to do so by establishing a database of the rat islet proteins and by identifying on high-resolution two-dimensional (2-D) gels proteins undergoing changes in expression levels following IL-1 exposure [43]. The database contains >2000 proteins. However, by visual analysis only 33 of them are influenced by IL-1 exposure: a few are induced de novo, e.g., iNOS, many are upregulated in expression, and a few are down-regulated. Currently, we are in the process of identifying these proteins either by (a) microsequencing proteins in spots cut out of preparatory 2-D gels (63 000 islets $\sim$ 200 μg protein), or (b) mass spectrometry [44], matrix-assisted laser desorption/ionization (MALDI) spectras for screening followed by database search for homologies. Of 13 proteins studied so far, positive identifications were obtained in four cases, and sequence information enough for subsequent cloning in an additional three, proving that this strategy may eventually identify all the proteins involved in the energy and protein synthesis requiring death process of the beta-cell. Together the functions of these proteins will probably help us to understand the cytokine-induced non-antigen-driven and non-lymphocyte-mediated initiation phase of beta-cell destruction.

Obviously, one (or more?) of these de novo synthesized islet protein(s), or the putative new "non-self" post-translatory free radical-induced immunogenic epitopes on beta-cell proteins, may turn out to be the primary driving antigen(s) responsible for the lymphocyte-mediated amplification and perpetuation phase of the IDDM pathogenesis. Systems to test for this, e.g. T-cell reactivity assays measuring Th1 immune responses, are in demand to solve this important issue, since this may provide some of the elements needed for the design of effective strategies aimed at the prevention and cure of IDDM. These observations and ideas are all compatible with our current model of the pathogenesis of IDDM.

The model predicts that anything from the external or internal environment which can destroy a beta-cell (nutrients, viruses, chemicals, IL-1) will lead to the release of beta-cell proteins.

These proteins will be taken up by residing antigen-presenting cells (APC). MØ, MO, DC in the islets will be processed to antigenic peptides and as such be presented by major histocompatibility complex (MHC)-class II molecules on the cell surface.

This activates the APC to produce and secrete monokines (IL-1, TNF) and co-stimulatory signal(s) which, if T-helper lymphocytes with receptors specifically recognizing the antigenic peptide are present in the islet, induce the transcription of a series of lymphokine genes. One of these, IFN, will feedback-stimulate the antigen-presenting cells to increase expression of MHC-class II molecules and IL-1 and TNF. In addition, other cells of the MØ/MO/DC lineage present in the islet are also induced to secrete monokines. IL-1, potentiated by TNF and IFN, is cytotoxic to beta-cells through the induction of free radical formation in the islet.

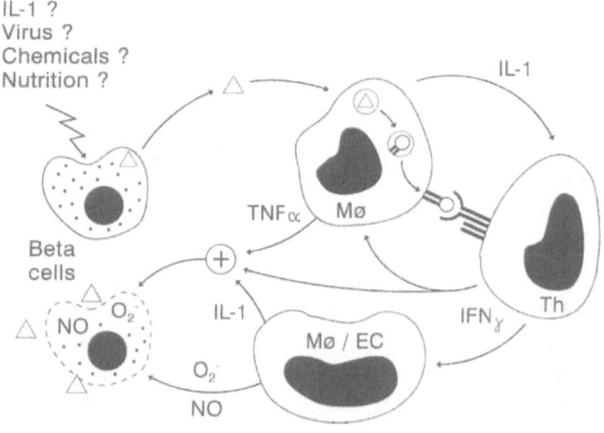

Fig. 1. The Copenhagen model of pathogenesis of insulin-dependent diabetes mellitus (IDDM). *TNFα*, tumour necrosis factor *α*; *IL-1*, interleukin-1; *IFNγ*, interferon-γ; *Th*, T-helper cell; *MØ*, mono-cyte; *EC*, endothelial cell

As part of the beta-cell destructive mechanism some beta-cell proteins are damaged ("denatured") by free radicals and in more antigenic forms (linear epitopes) presented to the immune system, thereby closing the loop in a self-perpetuating and self-limiting fashion.

The magnitude of beta-cell destruction will depend upon (a) the velocity of the feed-back circuit between the antigen-presenting and the T-helper lymphocyte, i.e. on the efficacy of antigen presentation–recognition, (b) the magnitude of cytokine production, and (c) on the capacity of beta-cell defence mechanisms during cytokine exposure. Each specific element of the pathogenetic process is under genetic control, i.e. IDDM is a polygenic disorder.

To focus on HLA-class II genes only is a much too reductionist approach and cannot fully explain the genetic background of IDDM.

The polygenetic nature of IDDM has been studied in two different ways: the *candidate gene* approach in which IDDM-associated polymorphisms of genes are identified as possible candidate genes from the pathogenetic model, and the *random marker approach*, which takes advantage of the most recent linkage maps of the human genome project. The first generation linkage map of human IDDM [45] provided some evidence for linkage of a rather large number of loci, some of which have been confirmed in independent data sets [46,47]. However, much larger IDDM sibpair family materials are required to identify finally all chromosomal segments containing genetic information involved in susceptibility to IDDM, and it is going to take considerable effort and time to identify the genes responsible for IDDM in these chromosomal segments. The candidate gene approach has identified IDDM-associated genes involved in antigen presentation and recognition, i.e., HLA-class II genes. This region represents the major locus responsible for IDDM and was obviously also identified as such by linkage studies [45]. Genes controlling the magnitude of cytokine production are likely to be involved. Certainly, IDDM-associated alleles of these genes encode for cytokine agonist hypersecretion and low circulating cytokine antagonist levels. The IL-1 and TNF systems are examples of this. The genes coding for IL-1 agonists, receptors and antagonists are located on the long arm of chromosome 2 in humans. Polymorphisms have been found in several of these genes and certain alleles of the IL1B, IL1RN and IL1RT1 gene polymorphisms are significantly associated with IDDM [48–50]. Although these gene polymorphisms have no influence on the qualitative activity of the peptide product, the possibility that the polymorphisms have quantitative consequences for the synthesis and secretion of the peptides has been investigated. Thus, in the case of IL-1β, there was a clear allele-dose effect of the IDDM-associated allele ILB*1 [48], in that stimulated monocyte IL-1β secretion was highest in IL-1B1 homozygous healthy individuals, intermediate in heterozygous individuals and lowest in those not carrying the IL1B1 allele. Similarly, IDDM patients homozygous for the IDDM-associated allele of the IL-1RN gene had lower circulating levels of the IL-1RN [50].

The genes for TNFα and β are located in the class III region of the MHC. The association between IDDM and certain alleles of the TNF loci are most likely due to linkage disequilibrium with class II susceptibility loci. This strong linkage disequilibrium makes it difficult to determine which particular locus may be the major IDDM susceptibility locus. Interestingly, in IDDM patients and controls selected to be identical with regard to HLA DR 3/4 there is an independent contribution of IDDM susceptibility by the TNFa2 microsatellite allele [51]. Notably, the TNFa2 microsatellite encodes for monocyte TNFα hypersecretion in healthy individuals. This means that polymorphic gene variants of this region with functional significance may be selected as preferred functional partners for particular HLA alleles with which they are in linkage disequilibrium, and thus they may contribute to IDDM susceptibility.

If certain subgroups of IDDM patients are studied, e.g. the 5%–10% which cannot form an HLA-class II heterodimer molecule on their cell surfaces, the contribution of non-HLA-class II genes is likely to be strong, as it might also be for cases of familial IDDM vs sporadic IDDM which, surprisingly, do not differ throughout the entire HLA-class I, III and I regions on chromosome 6p [33]. This was clearly demonstrated to be the case for the genes in the IL-1 gene cluster on chromosome 2g12-33. Genes controlling protective and deleterious mechanisms involved in beta-cell defence have, as yet, been less extensively studied. An IDDM-associated MnSOD gene polymorphism has been demonstrated [52] but all genes encoding the proteins undergoing changes in expression levels during cytokine exposure are candidate genes in this regard and should be studied for IDDM association and function when identified.

The model (see Fig. 1) has important implications, since it eliminates the need for the existence of specific environmental trigger(s), IDDM-specific genes or gene mutations, and targeted exposure of beta-cells to the cytotoxic action of monokines. Furthermore, it offers an explanation for the existing multitude of "autoantibodies" in newly diagnosed and pre-diabetic IDDM: proteins released during beta-cell destruction by free radicals may not be released in their native conformation, but rather in free radical modified ("denatured"), more antigenic forms. Against this background we postulate that the remarkable specificity of cellular destruction in the islets in IDDM is due to an inherent vulnerability of beta-cells to cytokine-induced free radical damage. In other words, the beta-cells die because they are beta-cells.

Putting together the jigsaw-puzzle of cytokine-induced protein changes and their genetic control should one day provide the full picture of the pathogenesis of IDDM at the molecular level and thereby, hopefully, the basis for its prevention and cure. The interplay between the environment, the immune response to the environment and the beta-cell response to the immune system is what makes up the pathogenesis of IDDM. This was probably already realized by Albert E. Renold in the early 1960s, since this way of thinking was probably behind his immunization experiments with insulin [53]. He would undoubtedly have liked to witness today's technically refined experimental approaches to the problem. And in his always supportive and encouraging way he would have convinced us that pursuing this type of research is still worthwhile.

References

1. Opie EL (1901) On the relation of chronic interstitial pancreatitis to the islands of Langerhans and to diabetes mellitus. J Exp Med 5: 397–428
2. Gepts W (1965) Pathologic anatomy of the pancreas in juvenile diabetes mellitus. Diabetes 14: 619–633
3. Roitt M, Doniach D (1958) Human autoimmune thyroiditis. Lancet II: 1027–1033

4. Anderson JR, Goudic RB, Gray KG, Timbury GC (1957) Autoantibodies in Addison's disease. Lancet I: 1123–1124
5. Nerup J, Bendixen G (1969) Antiadrenal cellular hypersensitivity in Addison's disease. II. Correlation with clinical and serological findings. Clin Exp Immunol 5: 355–363
6. Nerup J, Andersen OO, Bendixen G et al (1971) Antipancreatic cellular hypersensitivity in diabetes mellitus. Diabetes 20: 424–427
7. Bottazzo GF, Florin-Christensen A, Doniach D (1974) Islet cell antibodies in diabetes mellitus with autoimmune polyendocrine deficiencies. Lancet II: 1279–1282
8. Nerup J, Platz P, Andersen OO et al (1974) HLA antigens and diabetes mellitus. Lancet II: 864–866
9. Thomsen M, Platz P, Andersen OO et al (1975) MLC typing in juvenile diabetes mellitus and idiopathic Addison's disease. Transplant Rev 22: 125–147
10. Gamble DR (1976) A possible virus etiology for juvenile diabetes. In: Creutzfeldt W, Koebberling J, Neel JV (eds) The genetics of diabetes mellitus. Springer Verlag, Berlin Heidelberg New York, pp 95–105
11. Bækkeskov S, Aanstoot HJ, Christgau S et al (1990) Identification of the 64K autoantigen in insulin-dependent diabetes as the GABA-synthesizing enzyme glutamic acid decarboxylase. Nature 347, 151–156
12. Bækkeskov S, Nielsen JH, Marner B et al (1982) Autoantibodies in newly diagnosed diabetic children immunoprecipitate human pancreatic islet proteins. Nature 298: 167–169
13. Todd JA, Bell JI, McDevitt HO (1987) HLA-DQβ gene contributes to susceptibility and resistance to insulin-dependent diabetes mellitus. Nature 329: 599–604
14. Haskins K, McDuffie M (1990) Acceleration of diabetes in young NOD mice with a CD4[+] islet specific T cell clone. Science 249: 1433–1436
15. Boitard C, Larger E, Timsit J et al (1994) IDDM: An islet or an immune disease? Diabetologia 37[Suppl 2]: S90–S98
16. Mandrup-Poulsen T, Bendtzen K, Nerup J et al (1986) Affinity purified human interleukin 1 is cytotoxic to isolated islets of Langerhans. Diabetologia 29: 63–67
17. Mandrup-Poulsen T, Bendtzen K, Dinarello CA, Nerup J (1987) Human tumor necrosis factor potentiates human interleukin 1-mediated rat pancreatic β-cytotoxicity. J Immunol 39: 4077–4082
18. Romagnani S (1992) Induction of Th1 and Th2 responses: a key role for the "natural" immune response? Immunol Today 13: 379–381
19. Liblau RS, Singer SM, McDevitt HO (1995) Th1 and Th2 CD4+ T-cells in the pathogenesis of organ-specific autoimmune diseases. Immunol Today 16: 34–38
20. Hanenberg H, Kolb-Backhofen V, Kantwerk-Funke G, Kolb H (1989) Macrophage infiltration precedes and is prerequisite for lymphocytic insulitis in pancreatic islets of prediabetic BB rats. Diabetologia 32: 126–134
21. O'Reilly LA, Hutchings PR, Crocker PR et al (1991) Characterization of pancreatic islet cell infiltrates in NOD mice: effect of cell transfer and transgene expression. Eur J Immunol 21: 1171–1180
22. Mandrup-Poulsen T, Bentzen K, Nielsen JH et al (1985) Cytokines cause structural and functional damage to isolated islets of Langerhans. Allergy 40: 424–429
23. Jiang Z, Wada BA (1991) Cytokine gene expression in the islets of the diabetic Biobreeding/Worcester rat. J Immunol 146: 2990–94
24. Held W, MacDonald HR, Weismann IL et al (1990) Genes encoding tumor necrosis factor α and granzyme A are expressed during development of autoimmune diabetes. Proc Natl Acad Sci USA 87: 2239–2243
25. Oschilewski V, Kiesel U, Kolb H (1985) Administration of silica prevents diabetes in BB-rats. Diabetes 34: 197–199
26. Lee KU, Amano K, Yoon JW (1988) Evidence for initial involvement of macrophages in development of insulitis in NOD mice. Diabetes 37: 989–991
27. Kaufman DL, Claire-Salzler M, Tian J et al (1994) Spontaneous loss of T-cell tolerance to glutamic acid decarboxylase in immune insulin-dependent diabetes. Nature 366: 69–72
28. Hagopian WA, Karlsen AE, Gottsätter A et al (1993) Quantitative assay using recombinant human islet glutamic acid decarboxylase (GAD65) shows that 64K autoantibody positivity at onset predicts diabetes type. J Clin Invest 91: 368–374

29. Rambrand T, Pociot F, Knip M et al (1996) Autoantibodies to glutamic acid de-carboxylase and IA-2 are two independent markers of IDDM with distinct HLA associations. (submitted)

30. Petersen JS, Mackay P, Plesner A et al (1995) Tolerance induction to GAD65 or BSA does not affect the onset of diabetes in BB-rats. Diabetologia 38[Suppl 1]: A98

31. Plesner A, Worsaae A, Michelsen B et al (1994) High levels of T-cell responses to GAD65 in the NOD mouse compared to non-diabetic strains of mice. Diabetes 37[Suppl 1]: A92

32. Nerup J, Mandrup-Poulsen T, Helqvists et al (1994) On the pathogenesis of IDDM. Diabetologia 37[Suppl 2]: 82–89

33. Pociot F, Rønningen KS, Bergholdt R et al (1994) Genetic susceptibility markers in Danish patients with type 1 (insulin-dependent) diabetes: evidence for polygenicity in man. Autoimmunity 19: 169–178

34. Zamani M, Pociot F, Spaepen M et al (1996) Linkage and association of the HLA gene complex with IDDM in 81 Danish families. Strong linkage between DRβ1Lys 71+ and IDDM. Hum Genet 98: 491–496

35. Asayama K, Kooy NW, Burr IM (1986) Effect of vitamin E deficiency and selenium deficiency on insulin secretory reserve and free radical scavenging systems in islets: decrease of manganosuperoxide dismutase. J Clin Lab Med 107: 459–464

36. Sumoski W, Baquerizo H, Rabinovitch A (1989) Oxygen free radical scavengers protect rat islet cells from damage by cytokines. Diabetologia 32: 792–796

37. Helqvist S, Polla BS, Johannesen J, Nerup J (1991) Heat shock protein induction in rat pancreatic islets by recombinant human interleukin β. Diabetologia 34: 150–156

38. Kröncke KD, Funda J, Besschick B, Kolb-Bachoften V (1991) Macrophage cytotoxicity towards isolated rat islet cells: neither lysis nor its protection by nicotinamide is beta-cell specific. Diabetologia 34: 232–238

39. Corbett JA, Wang JL, Sweetland MA et al (1992) IL-1β induces the formation of nitric oxide by β-cells purified from rodent islets of Langerhans: evidence for the β-cell as a source and site of action of nitric oxide. J Clin Invest 90: 2384–2391

40. Delaney CA, Green MHL, Lowe JE, Green IC (1993) Endogenous nitric oxide induced by interleukin-1β in rat islets of Langerhans and Hit-T15 cells causes significant DNA damage as measured by the "comet" assay. FEBS Lett 333: 291–295

41. Albina JE, Cui S, Mateo RB, Reichner JS (1993) Nitric oxide-mediated apoptosis in murine peritoneal macrophages. J Immunol 150: 5080–5085

42. Borg LAH, Sandler S, Eizirik DL. Interleukin-1β increases the activity of superoxide dismutase in rat pancreatic islets. Diabetologia 34[Suppl 2]: A8

43. Andersen HU, Mose Larsen PM, Fey SJ et al (1995) Two-dimensional gel electrophoresis of rat islet proteins. Interleukin-1β-induced changes on protein expression are reduced by L-arginine depletion and nicotinamide. Diabetes 44: 400–407

44. Rasmussen HH, Mortz E, Mann M et al (1995) Identification of transformation sensitive proteins recorded in human two dimensional gel protein databases by mass-spectrometric peptide mapping and in combination with microsequencing. Electrophoresis 15: 406–416

45. Davies JL, Kawaguchi Y, Bennett ST et al (1994) A genome-wide search for human susceptibility genes. Nature 371: 130–136

46. Bennett ST, Lucassen AM, Gouch SCL et al (1995) Susceptibility to human type 1 diabetes at IDDM 2 is determined by tandem repeat variation at the insulin gene minisatellite locus. Nature Genetics 9: 284–292

47. Copeman JB, Hearne C, Cornall RJ et al (1995) Fine localisation of a type 1 diabetes susceptibility gene (IDDM 7) to human chromosome 2q by linkage disequilibrium mapping. Nature Genetics 9: 80–85

48. Pociot F, Mølvig J, Wogensen L et al (1992) A Taq1 polymorphism in the human interleukin-1β (IL-1β) gene correlated with IL-1β secretion in vitro. Eur J Clin Invest 22: 396–402

49. Bergholdt R, Karlsen AE, Johannesen J et al (1995) Characterization of polymorphisms of an interleukin-1 receptor type 1 promotor region (P2) and their relation to insulin-dependent diabetes mellitus (IDDM). Cytokine 7: 727–733

50. Mandrup-Poulsen T, Pociot F, Mølvig J et al (1994) Monokine antagonism is reduced in patients with IDDM. Diabetes 43: 1242–1247

51. Pociot F, Briant L, Jongeneel CV et al (1993) Association of tumor necrosis factor (TNF) and class II MHC alleles with the secretion of TNFα and TNFβ by human mononuclear cells: a possible link to insulin-dependent diabetes mellitus. Eur J Immunol 23: 224–231

52. Pociot F, Lorenzen T, Nerup J (1993) A manganese superoxide dismutase (SOD2) gene polymorphism in insulin-dependent diabetes mellitus. Dis Markers 11: 267–274

53. Renold AE, Soeldner JS, Steinke J (1964) Immunological studies with homologous and heterologous insulin in the cow. In: Cameron MP, O'Connor M (eds) Ciba Foundation colloquia on endocrinology, vol 15. Churchill, London, pp 122–134

Selected References from the Bibliography of Albert E. Renold, M.D.

1. Bernhard K, Favarger M, Renold A, Spühler O (1947) Beiträge zur Alloxan-Glucosurie. Helv Chim Acta 30: 1666–1672
2. Renold AE (1948) Der Alloxan-Diabetes. Thesis. Leeman, Zurich, 45pp
3. Marble A, Renold AE (1949) Atrophy and hypertrophy of subcutaneous fat due to insulin. Trans Assoc Am Physicians 62: 219–223
5. Renold AE, Marble A, Fawcett DW (1950) Action of insulin on deposition of glycogen and storage of fat in adipose tissue. Endocrinology 46: 55–56
10. Renold AE, Forsham PH, Maisterrena J, Thorn GW (1951) Intravenously administered ACTH. N Engl J Med 244: 796–798
12. Thorn GW, Renold AE, Wilson DL, Frawley TF, Jenkins D, Garcia-Reyes J, Forsham PH (1951) Clinical studies on the activity of orally administered cortisone. N Engl J Med 245: 549–555
15. Renold AE, Teng CT, Nesbett FB, Hastings AB (1953) Studies on carbohydrate metabolism in rat liver slices: II. The effect of fasting and of hormonal deficiencies. J Biol Chem 204: 533–546
16. Hastings AB, Renold AE, Teng CT (1953) Cationic and hormonal influences on carbohydrate metabolism of rat liver in vitro. Trans Assoc Am Physicians 66: 129–135
18. Renold AE, Hastings AB, Nesbett FG (1954) Studies on carbohydrate metabolism in rat liver slices: III. Utilization of glucose and fructose by liver of normal and diabetic animals. J Biol Chem 209: 687–696
21. Renold AE, Hastings AB, Nesbett FB, Ashmore J (1955) Studies on carbohydrate metabolism in rat liver slices: IV. Biochemical sequence of events after insulin administration. J Biol Chem 213: 135–146
24. Ashmore J, Renold AE, Nesbett FB, Hastings AB (1955) Studies on carbohydrate metabolism in rat liver slices: V. Glycerol metabolism in relation to other substrates in normal and diabetic tissue. J Biol Chem 215: 153–161
27. Renold AE, Abu Haydar N, Reddy WJ, Goldfien A, St. Marc JR, Laidlaw JC (1955) Biological effects of fluorinated derivatives of hydrocortisone and progesterone in man. Ann NY Acad Sci 61: 582–590
32. Ashmore J, Hastings AB, Nesbett FB, Renold AE (1956) Studies on carbohydrate metabolism in rat liver slices: VI. Hormonal factors influencing glucose-6-phosphatase. J Biol Chem 218: 77–88
36. Renold AE, Ashmore J, Hastings AB (1956) The regulation of carbohydrate metabolism in isolated tissues. Vitam Horm 14: 139–185
41. Renold AE, Winegrad AI, Froesch ER, Thorn GW (1956) Studies on the site of action of the arylsulfonylureas in man. Metabolism 5: 757–767
44. Renold AE, Crabbé J, Hernando L, Nelson DH, Ross EJ, Emerson K Jr, Thorn GW (1957) Inhibition of aldosterone secretion by amphenone in man. N Engl J Med 256: 16–21
50. Froesch ER, Ganong WF, Selenkow HA, Goodale W, Renold AE, Thorn GW (1957) Hyperglycemic effect without anabolic effect of beef growth hormome in man. Diabetes 6: 515–522
53. Hernando L, Crabbé J, Ross EJ, Reddy WJ, Renold AE, Nelson DH, Thorn GW (1957) Clinical experience with a physico-chemical method for the estimation of aldosterone in urine. Metabolism 6: 518–534
60. Ashmore J, Cahill GF Jr, Hillman R, Renold AE (1958) Adrenal cortical regulation of hepatic glucose metabolism. Endocrinology 62: 621–626

61. Martin DB, Renold AE, Dagenais YM (1958) An assay for insulin-like activity using rat adipose tissue. Lancet II: 76–77

62. Froesch ER, Ashmore J, Renold AE (1958) Comparison of renal and hepatic effects of fasting, cortisone administration and glucose infusion in normal and adrenalectomized rats. Endocrinology 62: 614–620

63. Winegrad AI, Renold AE (1958) Studies on rat adipose tissue in vitro. I. Effects of insulin on the metabolism of glucose, pyruvate, and acetate. J Biol Chem 233: 267–271

64. Winegrad AI, Renold AE (1958) Studies on rat adipose tissue in vitro. II. Effects of insulin on the metabolism of specifically labeled glucose. J Biol Chem 233: 273–276

65. Froesch ER, Winegrad AI, Renold AE, Thorn GW (1958) Mechanism of the glucosuria produced by the administration of steroids with glucocorticoid activity. J Clin Invest 37: 524–532

73. Renold AE, Zahnd GR, Jeanrenaud B, Boshell BR (1959) Some effects of tolbutamide and chlorpropamide in vitro. NY Acad Sci 75: 490–498

75. Felber JP, Renold AE, Zahnd GR (1959) The comparative metabolism of glucose, fructose, galactose and sorbitol in normal subjects and in disease states. Modern problems in pediatrics. 4: Carbohydrate metabolism in children. Karger, Basel, pp 467–489

80. Cahill GF Jr, Leboeuf B, Renold AE (1959) Studies on rat adipose tissue in vitro. III. Synthesis of glycogen and glyceride-glycerol. J Biol Chem 234: 2540–2543

82. Jeanrenaud B, Renold AE (1959) Studies on rat adipose tissue in vitro. IV. Metabolic patterns produced in rat adipose tissue by varying insulin glucose concentrations independently from each other. J Biol Chem 234: 3082–3087

87. Boshell BR, Zahnd GR, Renold AE (1960) An effect of tolbutamide on ketogenesis, in vivo and in vitro. Metabolism 9: 21–29

88. Bally P, Cahill GF Jr, Leboeuf B, Renold AE (1960) Studies on rat adipose tissue in vitro. V. Effects of glucose and insulin on the metabolism of palmitate-1-C^{14}. J Biol Chem 235: 333–336

91. Schwartz R, Schafer IA, Renold AE (1960) Generalized lipoatrophy, hepatic cirrhosis, disturbed carbohydrate metabolism and accelerated growth (lipoatrophic diabetes). Longitudinal observations and metabolic studies. Am J Med 28: 973–985

99. Cahill GF Jr, Leboeuf B, Renold AE (1960) Factors concerned with the regulation of fatty acid metabolism by adipose tissue. Am J Clin Nutr 8: 733–739

101. Zahnd G, Steinke J, Renold AE (1960) Early metabolic effects of human growth hormone. Proc Soc Exp Biol Med 105: 455–459

103. Jeanrenaud B, Renold AE (1960) Studies on rat adipose tissue in vitro. VII. Effects of adrenal cortical hormones. J Biol Chem 235: 2217–2223

105. Renold AE, Martin DB, Dagenais YM, Steinke J, Nickerson J, Sheps MC (1960) Measurement of small quantities of insulin-like activity using rat adipose tissue. I. A proposed procedure. J Clin Invest 39: 1487–1498

106. Sheps MC, Nickerson RJ, Dagenais YM, Steinke J, Martin DB, Renold AE (1960) Measurement of small quantities of insulin-like activity using rat adipose tissue. II. Evaluation of performance. J Clin Invest 39: 1499–1510

110. Wood FC Jr, Leboeuf B, Renold AE, Cahill GF Jr (1961) Metabolism of mannose and glucose by adipose tissue and liver slices from normal and alloxan-diabetic rats. J Biol Chem 236: 18–21

112. Christophe J, Jeanrenaud B, Mayer J, Renold AE (1961) Metabolism in vitro of adipose tissue in obese-hyperglycemic and goldthioglucose treated mice. I. Metabolism of glucose. J Biol Chem 236: 642–647

113. Christophe J, Jeanrenaud B, Mayer J, Renold AE (1961) Metabolism in vitro of adipose tissue in obese hyperglycemic and goldthioglucose treated mice. II. Metabolism of pyruvate and acetate. J Biol Chem 236: 648–652

116. Steinke J, Camerini R, Marble A, Renold AE (1961) Elevated levels of serum insulin-like activity (ILA) as measured with adipose tissue in early untreated diabetes and prediabetes. Metabolism 10: 707–711

117. Humbel RE, Renold AE, Herrera MG, Taylor KW (1961) Incorporation of glucose carbon into protein of the islets of Langerhans from toadfish (Opsanus tau). Endocrinology 69: 874–877

118. Taylor KW, Humbel RE, Steinke J, Renold AE (1961) The paper chromatography of insulins from ox pancreas, human pancreas and the isolated tissue of the N. American toadfish (Opsanus tau). Biochim Biophys Acta 54: 391–394

120. Leboeuf B, Renold AE, Cahill GF Jr (1962) Studies on rat adipose tissue in vitro. IX. Further effects of cortisol on glucose metabolism. J Biol Chem 237: 988–991
126. Steinke J, Sirek A, Lauris V, Lukens FDW, Renold AE (1962) Measurement of small quantities of insulin-like activity with rat adipose tissue. III. Persistence of serum insulin-like activity after pancreatectomy. J Clin Invest 41: 1699–1707
127. Humbel RE, Renold AE (1963) Studies on isolated islets of Langerhans (Brockmann bodies) of teleost fishes. I. Metabolic activity in vitro. Biochim Biophys Acta 74: 84–95
129. Steinke J, Soeldner JS, Renold AE (1963) Measurement of small quantities of insulin-like activity with rat adipose tissue. IV. Serum insulin-like activity and tumor insulin content in patients with functioning islet-cell tumors. J Clin Invest 42: 1322–1329
131. Evans JR, Opie LH, Renold AE (1963) Pyruvate metabolism in the perfused rat heart. Am J Physiol 205: 971–976
141. Renold AE, Soeldner JS, Steinke J (1964) Immunological studies with homologous and heterologous pancreatic insulin in the cow. In: Cameron MP, O'Connor M (eds) Ciba foundation colloquia on endocrinology, vol 15. Churchill, London, pp 122–134
148. Crofford OB, Renold AE (1965) Glucose uptake by incubated rat epididymal adipose tissue: rate limiting steps and site of insulin action. J Biol Chem 240: 14–21
149. Rafaelsen OJ, Lauris V, Renold AE (1965) Localized intraperitoneal action of insulin on rat diaphragm and epididymal adipose tissue in vivo. Diabetes 14: 19–26
151. Crofford OB, Renold AE (1965) Glucose uptake by incubated rat epididymal adipose tissue. Characteristics of the glucose transport system and action of insulin. J Biol Chem 240: 3237–3244
166. Gonet AE, Stauffacher W, Pictet R, Renold AE (1965) Obesity and diabetes mellitus with striking congenital hyperplasia of the islets of Langerhans in spiny mice (Acomys cahirinus). I. Histological findings and preliminary metabolic observations. Diabetologia 1: 162–171
168. Renold AE, Steinke J, Soeldner JS, Antoniades HN, Smith RE (1966) Immunological response to the prolonged administration of heterologous and homologous insulin in cattle. J Clin Invest 45: 702–713
169. Crofford OB, Stauffacher W, Jeanrenaud B, Renold AE (1966) Glucose transport in isolated fat cells. Procedures for measurement of the intracellular water space. Helv Physiol Acta 24: 45–57
172. Young DAB, Renold AE (1966) A fluorimetric procedure for the determination of ketone bodies in very small quantities of blood. Clin Chim Acta 13: 791–793
174. Vecchio D, Luyckx A, Zahnd GR, Renold AE (1966) Insulin release induced by glucagon in organ cultures of fetal rat pancreas. Metabolism 15: 577–681
181. Pictet R, Orci L, Gonet AE, Rouiller C, Renold AE (1967) Ultrastructural studies of the hyperplastic islets of Langerhans of spiny mice (Acomys cahirinus) before and during the development of hyperglycaemia. Diabetologia 3: 188–211
182. Stauffacher W, Lambert AE, Vecchio D, Renold AE (1967) Measurement of insulin activities in pancreas and serum of mice with spontaneous and induced obesity and hyperglycaemia, with considerations on the pathogenesis of the spontaneous syndrome. Diabetologia 3: 230–237
186. Ho RJ, Jeanrenaud B, Posternak T, Renold AE (1967) Insulin-like action of ouabain. II. Primary antilipolytic effect through inhibition of adenyl cyclase. Biochim Biophys Acta 144: 74–82
189. Letarte J, Renold AE (1967) Glucose metabolism in fat cells stimulated by insulin and dependent on sodium. Nature 215: 961–962
192. Junod A, Lambert AE, Orci L, Pictet R, Gonet AE, Renold AE (1967) Studies of the diabetogenic action of streptozotocin. Proc Soc Exp Biol Med 126: 201–205
195. Lambert A, Vecchio D, Gonet A, Jeanrenaud B, Renold AE (1967) Organ culture of fetal rat pancreas: effects of tolbutamide, glucagon and other substances. Excerpta Medica Foundation, Int Congress Series, Amsterdam 149: 61–82
196. Renold AE (1967) Spontaneous diabetes in laboratory animals. In: Internat symposium on laboratory animals; 5. Immunobiological Standards. Karger, Basel, pp 83–86
198. Letarte J, Renold AE (1968) Effects of cations upon glucose transport in isolated fat cells. In: Peeters H (ed) Protides of the biological fluids, vol 15. Elsevier, Amsterdam, pp 241–250

200. Orci L, Pictet R, Forssmann WG, Renold AE, Rouiller C (1968) Structural evidence for glucagon producing cells of the intestinal mucosa of the rat. Diabetologia 4: 56–67
207. Pictet R, Jeanrenaud B, Orci L, Renold AE, Rouiller C (1968) Cellules adipeuses in situ et isolées. Essai de fixation pour la microscopie électronique. Z Ges Exp Med 148: 255–274
209. Orci L, Junod A, Pictet R, Renold AE, Rouiller C (1968) Granulolysis in A cells of endocrine pancreas in spontaneous and experimental diabetes in animals. J Cell Biol 39: 462–466
213. Renold AE (1968) Spontaneous diabetes and/or obesity in laboratory rodents. In: Levine R, Luft R (eds) Advances in metabolic disorders. Academic, New York, 3: 49–84
216. Letarte J, Renold AE (1969) Ionic effects on glucose transport and metabolism by isolated mouse fat cells incubated with or without insulin. I. Lack of effect of medium Ca^{2+}, Mg^{2+} or PO_4^{3-}. Biochim Biophys Acta 183: 350–365
218. Letarte J, Renold AE (1969) Ionic effects on glucose transport and metabolism by isolated mouse fat cells incubated with or without insulin. III. Effects of replacement of Na^+. Biochim Biophys Acta 183: 366–374
219. Orci L, Lambert AE, Rouiller Ch, Renold AE, Samols E (1969) Evidence for the presence of A-cells in the endocrine fetal pancreas of the rat. Horm Metab Res 1: 108–110
221. Forssmann WG, Orci L, Pictet R, Renold AE, Rouiller C (1969) The endocrine cells in the epithelium of the gastrointestinal mucosa of the rat. J Cell Biol 40: 692–715
222. Lambert AE, Junod A, Stauffacher W, Jeanrenaud B, Renold AE (1969) Organ culture of fetal rat pancreas. I. Insulin release induced by caffeine and by sugars and some derivatives. Biochim Biophys Acta 184: 529–539
223. Lambert AE, Jeanrenaud B, Junod A, Renold AE (1969) Organ culture of fetal rat pancreas. II. Insulin release induced by amino and organic acids, by hormonal peptides, by cationic alterations of the medium and by other agents. Biochim Biophys Acta 184: 540–553
225. Burr IM, Stauffacher W, Balant L, Renold AE, Grodsky GM (1969) Dynamic aspects of proinsulin release from perifused rat pancreas. Lancet II: 882–883
227. Lambert AE, Orci L, Kanazawa Y, Renold AE (1969) Biosynthesis and release of insulin in organ cultures of fetal rat pancreas. In: Loubatières A, Renold AE (eds) Pharmacokinetics and mode of action of oral hypoglycemic agents. Acta Diabetica Latina 6 [Suppl 1]: 505–537
231. Stauffacher W, Renold AE (1969) Effect of insulin in vivo on diaphragm and adipose tissue of obese mice. Am J Physiol 216: 98–105
232. Junod A, Lambert AE, Stauffacher W, Renold AE (1969) Diabetogenic action of streptozotocin: Relationship of dose to metabolic response. J Clin Invest 48: 2129–2139
256. Burr IM, Balant L, Stauffacher W, Renold AE (1971) Adrenergic modification of glucose-induced biphasic insulin release from perifused rat pancreas. Eur J Clin Invest 1: 216–224
257. Balant L, Burr IM, Stauffacher W, Cameron DP, Buenzi HF, Humbel RE, Renold AE (1971) Insulin of spiny mice (Acomys cahirinus) – partial characterization and evidence for two insulins. Endocrinology 88: 517–521
260. Orci L, Stauffacher W, Rufener C, Lambert AE, Rouiller C, Renold AE (1971) Acid phosphatase activity in secretory granules of pancreatic b-cells of normal rats. Diabetes 20: 385–388
263. Assal J-P, Levrat R, Stauffacher W, Renold AE (1971) Metabolic consequences of portacaval shunting in the rat: effects on glucose tolerance and serum immunoreactive insulin response. Metabolism 20: 850–858
264. Burr IM, Marliss EB, Stauffacher W, Renold AE (1971) Diazoxide effects on biphasic insulin release: "adrenergic" suppression and enhancement in the perifused rat pancreas. J Clin Invest 50: 1444–1450
266. Renold AE (1971) Research and physician training in a European university medical school. In: Condliffe PG, Furnia AH (eds) Reform of medical education. The role of research in medical education. Fogarty International Center Proceedings No. 7, Bethesda, pp 167–181
268. Burr IM, Marliss EB, Stauffacher W, Renold AE (1971) Differential effect of ouabain on glucose-induced biphasic insulin release in vitro. Am J Physiol 221: 943–947
272. Orci L, Lambert AE, Kanazawa Y, Amherdt M, Rouiller C, Renold AE (1971) Morphological and biochemical studies of B-cells of fetal rat endocrine pancreas in organ culture. Evidence for (pro)insulin biosynthesis. J Cell Biol 50: 565–582

277. Lambert AE, Blondel B, Kanazawa Y, Orci L, Renold AE (1972) Monolayer cell culture of neonatal rat pancreas: light microscopy and evidence for immunoreactive insulin synthesis and release. Endocrinology 90: 239–248

281. Renold AE, with the advice and collaboration of Orci L, Kanazawa Y, Marliss EB, Stauffacher W, Blondel B, Amherdt M, Lambert AE, Seydoux J, Girardier L, Porte D (1972) The b-cell and its responses: summarizing remarks and some contributions from Geneva. Diabetes 21 [Suppl 2]: 619–631

283. Cameron DP, Stauffacher W, Orci L, Amherdt M, Renold AE (1972) Defective immunoreactive insulin secretion in the Acomys cahirinus. Diabetes 21: 1060–1071

286. Marliss EB, Wollheim CB, Blondel B, Orci L, Lambert AE, Stauffacher W, Like AA, Renold AE (1973) Insulin and glucagon release from monolayer cell cultures of pancreas from newborn rats. Eur J Clin Invest 3: 16–26

287. Orci L, Amherdt M, Malaisse-Lagae F, Rouiller C, Renold AE (1973) Insulin release by emiocytosis: demonstration with freeze-etching technique. Science 179: 82–83

289. Marliss EB, Girardier L, Seydoux J, Wollheim CB, Kanazawa Y, Orci L, Renold AE, Porte D Jr (1973) Glucagon release induced by pancreatic nerve stimulation in the dog. J Clin Invest 52: 1246–1259

290. Orci L, Like AA, Amherdt M, Blondel B, Kanazawa Y, Marliss EB, Lambert AE, Wollheim CB, Renold AE (1973) Monolayer cell culture of neonatal rat pancreas: an ultra-structural and biochemical study of functioning endocrine cells. J Ultrastruct Res 43: 270–297

291. Orci L, Malaisse-Lagae F, Ravazzola M, Amherdt M, Renold AE (1973) Exocytosis-endocytosis coupling in the pancreatic beta cell. Science 181: 561–562

294. Orci L, Amherdt M, Henquin JC, Lambert AE, Unger RH, Renold AE (1973) Pronase effect on pancreatic beta cell secretion and morphology. Science 180: 647–649

295. Cameron DP. Stauffacher W, Amherdt M, Orci L, Renold AE (1973) Kinetics of immunoreactive insulin release in obese hyperglycemic laboratory rodents. Endocrinology 92: 257–264

296. Orci L, Perrelet A, Ravazzola M, Malaisse-Lagae F, Renold AE (1973) A specialized membrane junction between nerve endings and B-cells in islets of Langerhans. Eur J Clin Invest 3: 443–445

299. Kikuchi M, Rabinovitch A, Blackard WG, Renold AE (1974) Perifusion of pancreas fragments. A system for the study of dynamic aspects of insulin secretion. Diabetes 23: 550–559

300. Amherdt M, Harris V, Renold AE, Orci L, Unger RH (1974) Hepatic autophagy in uncontrolled experimental diabetes and its relationships to insulin and glucagon. J Clin Invest 54: 188–193

305. Gutzeit A, Rabinovitch A, Karakash C, Stauffacher W, Renold AE, Cerasi E (1974) Evidence for decreased sensitivity to glucose of isolated islets from spiny mice (Acomys cahirinus). Diabetologia 10: 661–665

306. Gutzeit A, Rabinovitch A, Studer PP, Trueheart PA, Cerasi E, Renold AE (1974) Decreased intravenous glucose tolerance and low plasma insulin response in spiny mice (Acomys cahirinus). Diabetologia 10: 667–670

309. Wollheim CB, Blondel B, Trueheart PA, Renold AE, Sharp GWG (1975) Calcium-induced insulin release in monolayer culture of the endocrine pancreas. Studies with ionophore A 23187. J Biol Chem 250: 1354–1360

310. Cuendet GS, Loten EG, Cameron DP, Renold AE, Marliss EB (1975) Hormone-substrate response to total fasting in lean and obese mice. Am J Physiol 228: 276–283

311. Blackard WG, Kikuchi M, Rabinovitch A, Renold AE (1975) An effect of hypoosmolarity on insulin release in vitro. Am J Physiol 228: 706–713

312. Sharp GWG, Wollheim CB, Muller WA, Gutzeit A, Trueheart PA, Blondel B, Orci L, Renold AE (1975) Studies on the mechanism of insulin release. Fed Proc 34: 1537–1548

314. Rabinovitch A, Gutzeit A, Grill V, Kikuchi M, Cerasi E, Renold AE (1975) Defective early phase insulin release in perifused isolated pancreatic islets of spiny mice (Acomys cahirinus). Diabetologia 11: 457–465

315. Orci L, Malaisse-Lagae F, Ravazzola M, Rouiller D, Renold AE, Perrelet A, Unger R (1975) A morphological basis for intracellular communication between alpha- and beta-cells in the endocrine pancreas. J Clin Invest 56: 1066–1070

317. Rabinovitch A, Gutzeit A, Renold AE, Cerasi E (1975) Insulin secretion in the spiny mouse (Acomys cahirinus). Dose and time kinetic studies with glucose in vivo and in vitro. Diabetes 24: 1094–1100

318. Kramp RC, Congdon CC, Gutzeit A, Renold AE (1975) Subcutaneous transplantation of pancreatic islet tissue in diabetic mice. Transplant Proc 7[Suppl 1]: 735–738

323. Wollheim CB, Blondel B, Renold AE, Sharp GWG (1976) Calcium induced glucagon release in monolayer culture of the endocrine pancreas. Studies with ionophore A 23187. Diabetologia 12: 287–294

324. Maldonato A, Trueheart PA, Renold AE, Sharp GWG (1976) Effects of streptozotocin in vitro on proinsulin biosynthesis, insulin release and ATP content of isolated rat islets of Langerhans. Diabetologia 12: 471–481

325. Rabinovitch A, Renold AE, Cerasi E (1976) Decreased cyclic AMP and insulin responses to glucose in pancreatic islets of diabetic Chinese hamsters. Diabetologia 23: 581–587

328. Cuendet GS, Loten EG, Jeanrenaud B, Renold AE (1976) Decreased basal, noninsulin-stimulated glucose uptake and metabolism by skeletal soleus muscle isolated from obese-hyperglycemic (obob) mice. J Clin Invest 58: 1078–1088

330. Kramp RC, Cuche R, Renold AE with the technical assistance of Gonet JC (1976) Subcutaneous, isogeneic transplantation of either duct-ligated pancreas or isolated islets in streptozotocin diabetic mice. Endocrinology 99: 1161–1167

331. Muller WA, Amherdt ML, Vachon C, Renold AE (1976) Phosphorylation reactions in islets of Langerhans. J Physiol Paris 72: 711–720

336. Wollheim CB, Blondel B, Renold AE, Sharp GWG (1977) Somatostatin inhibition of pancreatic glucagon release from monolayer cultures and interactions with calcium. Endocrinology 101: 911–919

337. Wollheim CB, Kikuchi M, Renold AE, Sharp GWG (1977) Somatostatin- and epinephrine-induced modification of $^{45}Ca^{++}$ fluxes and insulin release in rat pancreatic islets maintained in tissue culture. J Clin Invest 60: 1165–1173

338. Maldonato A, Renold AE, Sharp GWG, Cerasi E (1977) Glucose-induces proinsulin biosynthesis. Role of islet cyclic AMP. Diabetes 26: 538–544

343. Kikuchi M, Wollheim CB, Cuendet GS, Renold AE, Sharp GWG (1978) Studies on the dual effects of glucose on Ca^{++} efflux from isolated rat islets. Endocrinology 102: 1339–1349

344. Berger M, Halban PA, Muller WA, Offord RE, Renold AE, Vranic M (1978) Mobilization of subcutaneously injected tritiated insulin in rats: effects of muscular exercise. Diabetologia 15: 133–140

345. Müller WA, Girardier L, Seydoux J, Berger M, Renold AE, Vranic M (1978) Extrapancreatic glucagon and glucagon-like immunoreactivity in depancreatized dogs. A quantitative assessment of secretion rates and anatomical delineation of sources. J Clin Invest 62: 124–132

346. Wollheim CB, Kikuchi M, Renold AE, Sharp GWG (1978) The roles of intracellular and extracellular Ca^{++} in glucose-stimulated biphasic insulin release by rat islets. J Clin Invest 62: 451–458

348. Rabinovitch A, Cuendet GS, Sharp GWG, Renold AE, Mintz DH (1978) Relation of insulin release to cyclic AMP content in rat pancreatic islets maintained in tissue culture. Diabetes 27: 766–773

352. Niesor EJ, Wollheim CB, Mintz DH, Blondel B, Renold AE, Weil R (1979) Establishment of rat pancreatic endocrine cell lines by infection with Simian virus 40. Biochem J 178: 559–568

353. Siegel EG, Wollheim CB, Sharp GWG, Herberg L, Renold AE (1979) Defective calcium handling and insulin release in islets from diabetic Chinese hamsters. Biochem J 180: 233–236

356. Muller WA, Berger M, Suter P, Cüppers HJ, Reiter J, Wyss T, Berchtold P, Schmidt FH, Assal JP, Renold AE (1979) Glucagon immunoreactivities and amino acid profile in plasma of duodenopancreatectomized patients. J Clin Invest 63: 820–827

358. Gutzeit A, Renold AE, Cerasi E, Shafrir E (1979) Effect of diet-induced obesity on glucose and insulin tolerance of a rodent with a low insulin response (Acomys cahirinus). Diabetes 28: 777–784

362. Trimble ER, Vassutine I, Debray-Sachs M, Renold AE (1979) Long survival of pancreatic islets transplanted without immunosuppression across weak histocompatibility barriers. Transplant Proc 11: 1995–1999

363. Siegel EG, Wollheim CB, Kikuchi M, Renold AE, Sharp GWG (1980) Dependency of cyclic AMP-induced insulin release on intra- and extracellular calcium in rat islets of Langerhans. J Clin Invest 65: 233–241

366. Siegel EG, Trapp VE, Wollheim CB, Renold AE, Schmidt FH (1980) Beneficial effects of low-carbohydrate-high protein diets in long-term diabetic rats. Metabolism 29: 421–428

367. Trimble ER, Karakash C, Malaisse-Lagae F, Orci L, Renold AE (1980) Effects of intraportal islet transplantation of the transplanted tissue and the recipient pancreas. Diabetes 29: 341–347

370. Siegel EG, Wollheim CB, Sharp GWG, Herberg L, Renold AE (1980) Role of Ca^{2+} in impaired insulin release from islets of diabetic (C57BL/KsJ-db/db) mice. Am J Physiol 239: E132–E138

371. Siegel EG, Trimble ER, Berthoud H-R, Bereiter DA, Renold AE (1980) Effect of absence of early insulin response on oral glucose tolerance in islet-transplanted rats. Transplant Proc 12: 192–194

372. Meda P, Halban PA, Perrelet A, Renold AE, Orci L (1980) Gap junction development is correlated with insulin content in the pancreatic B cell. Science 209: 1026–1028

374. Siegel EG, Wollheim CB, Renold AE, Sharp GWG (1980) Evidence for the involvement of Na/Ca exchange in glucose-induced insulin release from rat pancreatic islets. J Clin Invest 66: 996–1003

375. Trimble ER, Herbert L, Renold AE (1980) Hypersecretion of pancreatic somatostatin in the obese Zucker rat. Effects of food restriction and age. Diabetes 29: 889–894

376. Siegel EG, Trimble ER, Renold AE, Berthoud HR (1980) Importance of preabsorptive insulin release on oral glucose tolerance: studies in pancreatic islet transplanted rats. Gut 21: 1002–1009

381. Philippe J, Halban PA, Gjinovci A, Duckworth WC, Estreicher J, Renold AE (1981) Increased clearance and degradation of [^{3}H]insulin in streptozotocin diabetic rats. J Clin Invest 67: 673–680

382. Gerber PP, Trimble ER, Wollheim CB, Renold AE, Miller RE (1981) Glucose and cyclic AMP as stimulators of somatostatin and insulin secretion from the isolated perfused rat pancreas. A quantitative study. Diabetes 30: 40–44

383. Ribes G, Siegel EG, Wollheim CB, Renold AE, Sharp GWG (1981) Rapid changes in calcium content of rat pancreatic islets in response to glucose. Diabetes 30: 52–55

384. Carpentier J-L, Halban PA, Renold AE, Orci L (1981) Internalization of ^{125}I-insulin by IM-9 cultured human lymphocytes: a comparison between A14-monoiodo-pork and biosynthetic human insulin. Diabetes Care 4: 220–222

388. Trimble ER, Renold AE (1981) Ventral and dorsal areas of rat pancreas: islet hormone content and secretion. Am J Physiol 240: E422–E427

390. Gerber PPG, Trimble ER, Wollheim CB, Renold AE (1981) Effect of insulin on glucose- and arginine-stimulated somatostatin secretion from the isolated perfused rat pancreas. Endocrinology 109: 279–283

391. Trimble ER, Gerber PPG, Renold AE (1981) Abnormalities of pancreatic somatostatin secretion corrected by in vivo insulin treatment of streptozotocin-diabetic rats. Diabetes 30: 865–867

394. Trimble ER, Berthoud HR, Siegel EG, Jeanrenaud B, Renold AE (1981) Importance of cholinergic innervation of the pancreas for glucose tolerance in the rat. Am J Physiol 241: E337–E341

396. Trimble ER, Gerber PG, Renold AE (1982) Effects of synthetic human pancreatic polypeptide, synthetic bovine pancreatic somatostatin and glucagon secretion in the rat. Diabetes 31: 178–181

397. Trimble ER, Halban PA, Wollheim CB, Renold AE (1982) Functional differences between rat islets of ventral and dorsal pancreatic origin. J Clin Invest 69: 405–412

399. Kaubisch N, Hammer R, Wollheim CB, Renold AE, Offord RE (1982) Specific receptors for sulfonylureas in brain and in a B-cell tumor of the rat. Biochem Pharmacol 32: 1171–1174

403. Masiello P, Wollheim CB, Janjic D, Gjinovci A, Blondel B, Praz G, Renold AE (1982) Stimulation of insulin release by glucose in a transplantable rat islet cell tumor. Endocrinology 111: 2091–2096

404. Masiello P, Wollheim CB, Blondel B, Renold AE (1983) Studies in vivo and in vitro on chemically-induced primary islet cell tumours and non-tumour endocrine pancreatic tisssue. Diabetologia 24: 30–37

405. Praz GA, Halban PA, Wollheim CB, Blondel B, Strauss AJ, Renold AE (1983) Regulation of immunoreactive insulin release from a rat cell line (RINm5F). Biochem J 210: 345–352

407. Halban PA, Renold AE (1983) Influence of glucose on insulin handling by rat islets in culture. A reflection of integrated changes in insulin biosynthesis, release, and intracellular degradation. Diabetes 32: 254–261

412. Prentki M, Renold AE (1983) Neutral amino acid transport in isolated rat pancreatic islets. J Biol Chem 258: 14239–14244

415. Halban PA, Amherdt M, Orci L, Renold AE (1984) Proinsulin modified by analogues of arginine and lysine is degraded rapidly in pancreatic B-cells. Biochem J 219: 91–97

416. Bruzzone R, Trimble ER, Gjinovci A, Renold AE (1984) Changes in content and secretion of pancreatic enzymes in the obese Zucker rat. Biochem J 219: 333–336

418. Weir GC, Halban PA, Meda P, Wollheim CB, Renold AE (1984) Dispersed adult rat pancreatic islet cells in culture: A, B, and D cell function. Metabolism 33: 447–453

422. Trimble ER, Weir GC, Gjinovci A, Assimacopoulos-Jeannet F, Benzi R, Renold AE (1984) Increased insulin responsiveness in vivo and in vitro consequent to induced hyperinsulinemia in the rat. Diabetes 33: 444–449

429. Pernet A, Trimble ER, Kuntschen F, Assal J-P, Hahn C, Renold AE (1985) Sulfonylureas in insulin-dependent (type I) diabetes: evidence for an extrapancreatic effect in vivo. J Clin Endocrinol Metab 61: 247–251

430. Trimble ER, Bruzzone R, Gjinovci A, Renold AE (1985) Activity of the insulo-acinar axis in the isolated perfused rat pancreas. Endocrinology 117: 1246–1252

432. Janjic D, Wollheim CB, Renold AE (1985) Calmodulin in various insulin secreting cell types. J Endocrinol Invest 8: 423–426

433. Iynedjian PB, Möebius G, Seitz HJ, Wollheim CB, Renold AE (1986) Tissue-specific expesssion of glucokinase: identification of the gene product in liver and pancreatic islets. Proc Natl Acad Sci USA 83: 1998–2001

434. Halban PA, Amherdt M, Orci L, Renold AE (1986) Tris(hydroxymethyl)aminomethane inhibits the synthesis and processing of proinsulin in isolated rat pancreatic islets without affecting release of insulin stores. Diabetes 35: 433–439

436. Bruzzone R, Trimble ER, Gjinovci A, Renold AE (1986) Differences in pancreatic enzyme release from ventral and dorsal areas of the rat pancreas. Am J Physiol 251: G56–G64

438. Iynedjian PB, Gjinovci A, Renold AE (1988) Stimulation by insulin of glucokinase gene transcription in liver of diabetic rats. J Biol Chem 263: 740–744

443. Dayer-Métroz M-D, Kimoto M, Izui S, Vassalli P, Renold AE (1988) Effect of helper and/or cytotoxic T-lymphocyte depletion on low-dose streptozotocin-induced diabetes in C57BL/6J mice. Diabetes 37: 1082–1089

444. Wilberz S, Herberg L, Renold AE (1988) Gangliosides in vivo reduce diabetes incidence in non obese diabetic mice. Diabetologia 31: 855–857

Subject Index

acetyl-CoA carboxylase   10, 16, 57
acromegaly   136
actin   5, 81, 110
adenylate cyclase   4, 50, 75
adhesion molecules   24
adipocytes   1–3, 110, 113, 115, 119
adipose tissue   133
anabolism   10–12, 129
animal models for diabetes
   BB rat   149, 151
   NOD mouse   149
   ob/ob-C57 mouse   18
   streptozotocin rat   112, 138
   transgenic mice, related to
      glucokinase   65
      GLUT-4   115
      insulin receptor   98
      IRS-1   98
   Zucker rat   113
L-arabinose   122
ATP   4, 54
   sensitive K$^+$ channel   70
autoantibodies   149
autoantigen   149
autocrine mechanisms   128, 136

β-cell   25, 45, 54, 69, 148
   aggregation   25
   destruction   149, 152
   lines   30, 49, 74
   membrane   79
   mitochondria   59
   pairs   33
   single   30–32, 71
   transfection   30, 45, 47
"big" IGF II   136
bis-mannose   116
   photolabelling   117

calcium (CA^{2+})   8, 59–62
   channels   70, 80
   flux   70
   ionophore   75
   oscillations   28

calmodulin   74
caveolae   8
CD4+ cells   149
CD8+ cells   149
cell
   coupling   29, 34–36
   osmolarity   17
   pH   15
   swelling   12–19
   transfection   47
cholecystokinin   73
chromaffin cells   77
citric acid cycle   59
clamp euglycaemic   138, 148
connexins   28
C-peptide   44, 76, 140
CRE (cyclic AMP responsive element)   50
cyclic AMP   8, 48, 51, 119
cytochalasin B   110, 115
cytokines   128, 153
cytoskeleton   80
cytotoxicity   149–150
   of monokines   156

delta cell   27
2-desoxy glucose   55
diacylglycerol   74
diazoxide   72
DNA, mitochondrial   69
DR3-alleles   148
DR4-alleles   148
dwarfism   131, 137

electron microscopy   29, 115
endocytosis
   glucose transporter   118
   insulin receptor   91
endosome fusion   77
energy expenditure   130, 141
epinephrine   50, 75
erythrocyte membrane   114, 116
exocytosis
   glucose transporter   118
   insulin   72

exocytosis, insulin (*Contd.*)
    regulation of   77

fasting   131, 134
FOR (oxygen-derived free radical)   153
free fatty acids   129, 142
fructose-6-P   57

GAD (glutamic acid decarboxylase)   149
galactose   56, 119
galanin   51, 75
gap junctions   24–25, 28–30, 36
gene
    duplication   44, 57, 129
    enhancer   44–45
    linkage maps   155
    polymorphism   155
    promoter   44–45
    regulation   45
glibenclamide   38
glucagon   6, 14, 27, 114
    gene expression   47
GLP-1 (glucagon-like peptide)   50, 73
glucose
    homeostasis   99, 108, 128, 131
    sensor   55
    transport   55
glucose transporter
    cloning   45
    isoforms   108, 115
    translocation   110
    turnover   113
glycogen
    muscle   130
    synthase in liver   19
    synthesis in liver   10, 16
glyconeogenesis   10
Golgi complex   69, 112
G-proteins (guanine nucleotide binding
        proteins)   4, 74, 118
growth   128
    hormone effects   133
        secretion   134
gut hormones   26, 73

haemolytic plaque assay, insulin   31
hepatocytes   10, 14, 18, 57
hep G2-cells   114
hexose, transferases   57
HIT-T15 cells   49, 76, 78
HLA-IDDM association   149
hyperinsulinaemia   170

IDDM   148–150
    familial   156
    polygenetic nature   155
    sporadic   156

γ-IFN (interferon)   154
insulin
    actions   89, 109, 127
    analogues   90, 127
    antibodies   150
    biosynthesis   48, 68
    exocytosis   69, 80
    gene
        regulation   46
        structure   45
    resistance   113, 115
    secretion
        biphasic   80
        pulsatile   72
insulin-like growth factors (IGFs)
    autocrine mechanisms   128, 136
    binding proteins   135
    mitogenic effects   129
    paracrine mechanisms   136
    primary structure   128
    receptor   90, 129
    recombinant (hrIGF I), clinical use   139
insulin receptor
    crosslinking   89
    signalling pathways   92
    subcellular trafficking   112
    substrates (IRS-1, IRS-2)   92
    subunits   89
insulitis   114
interleukins (IL-1, IL-2)   149
IP$_3$ (inositol 1,4,5-triphosphate)   12, 70,
        94
    kinase   18, 96
islet of Langerhans   24, 26, 72
isoproterenol   118

K$^+$ channels   14, 70, 72
ketogenesis   139

lipogenesis   10
lipolysis   10, 139
lipolytic hormones   3
liver   10, 12, 133, 135
lysosomes   112

magnesium (Mg^{2+})   6, 17
mannoheptulose   55
MAPK (mitogen-activated protein kinase)   17,
        20
    cascade   17, 95
membrane
    action potential   71
    depolarisation   28, 70
    fusion   70, 81
3-0-methylglucose   110, 118
microfilaments   81
mitochondria, metabolism   71

monocyte-macrophage-dendritic cell
    lineage   149
monokines   154
  cytotoxic action   156
muscle
  heart   129
  skeletal   110, 119
myosin   5

NAD(P)H fluorescence   71
neurotoxins   81
neurotransmitter secretion   70
NIDDM   38
NO (nitric oxide)   153
NOD mouse   149
NOS (nitric oxide synthetase)   153
NSF (N-ethyl maleimide sensitive factor)   80
NSILA (non-suppressible insulin-like
    activity)   132

ob/ob-C57 mouse   55
osmotic shock   18

pancreas, development   47, 51
paracrine, mechanisms   27, 136
patch clamp   72
PDGF (platelet-derived growth factor)   97
PEPCK (phosphoenol-pyruvate
    carboxykinase)   10, 12, 59
phospholipase C   70, 74
photolabelling   116
plasma membrane   2, 5, 24, 109
pleckstrin homology domain   93
preproinsulin   45
protein kinase A   50, 89
  B   23, 95
  C   73, 118
protein phosphatase   17
PTKs (protein tyrosine kinases)   89
pyruvate   55
  metabolism   59

Rab proteins   78

rabphilin   78
ras   93
RINm5F cells   74

scavengers   153
second messengers   70
secretory granules   69
  vesicles   81
    docking   84
    fusion   84
signal transduction   6, 74, 97
α SNAP (soluble NSF attachment protein)
    80
SNAP-25 (synaptosomal-associated protein)
    81
SNARE (α SNAP receptor)   82
SOD (superoxide dismutase)   153
somatostatin   27, 51
src homology domain   92
stimulus-secretion coupling   68
streptozotocin diabetic rat   112, 138
stress proteins   153
sulphonylureas   38, 70
synaptotagmin   81
syntaxin   84

T-cell reactivity   151
T-lymphocytes   149
α-TNF (tumour necrosis factor)   154
transcription factors   17, 46, 90
transducer   11, 15
transfection   30, 47
transgenic animals   45, 65, 98, 115
tubulin   5, 120
tumour hypoglycaemia   136

VAMP (vesicle-associated membrane
    protein)   81
video imaging   73

wortmannin   96

Zucker rat   113

Springer
and the
environment

At Springer we firmly believe that an international science publisher has a special obligation to the environment, and our corporate policies consistently reflect this conviction.

We also expect our business partners – paper mills, printers, packaging manufacturers, etc. – to commit themselves to using materials and production processes that do not harm the environment. The paper in this book is made from low- or no-chlorine pulp and is acid free, in conformance with international standards for paper permanency.

Springer